Family Game Night

Organize Epic Gaming Events for Friends and Family

(Games Challenges That Bring You and Your Partner Together Like Never Before)

George Sparks

Published By **Cathy Nedrow**

George Sparks

All Rights Reserved

Family Game Night: Organize Epic Gaming Events for Friends and Family (Games Challenges That Bring You and Your Partner Together Like Never Before)

ISBN 978-1-7751012-8-4

No part of this guidebook shall be reproduced in any form without permission in writing from the publisher except in the case of brief quotations embodied in critical articles or reviews.

Legal & Disclaimer

The information contained in this book is not designed to replace or take the place of any form of medicine or professional medical advice. The information in this book has been provided for educational & entertainment purposes only.

The information contained in this book has been compiled from sources deemed reliable, and it is accurate to the best of the Author's knowledge; however, the Author cannot guarantee its accuracy and validity and cannot be held liable for any errors or omissions. Changes are periodically made to this book. You must consult your doctor or get professional medical advice before using any of the suggested remedies, techniques, or information in this book.

Upon using the information contained in this book, you agree to hold harmless the Author from and against any damages, costs, and expenses, including any legal fees potentially resulting from the application of any of the information provided by this guide. This disclaimer applies to any damages or injury caused by the use and application, whether directly or indirectly, of any advice or information presented, whether for breach of contract, tort, negligence, personal injury, criminal intent, or under any other cause of action.

You agree to accept all risks of using the information presented inside this book. You need to consult a professional medical practitioner in order to ensure you are both able and healthy enough to participate in this program.

Table Of Contents

Chapter 1: Play Outside 1

Chapter 2: Rainy Day Indoor Games 20

Chapter 3: Group Games 41

Chapter 4: Solo Games........................... 59

Chapter 5: Educational Games 70

Chapter 6: World History To Remember 87

Chapter 7: What Type Of Whale Can Grow A Prolonged Tusk? 112

Chapter 8: What River Feeds Grand Canyon Of The Yellowstone? 128

Chapter 9: Partner Games..................... 145

Chapter 10: Family Golf 160

Chapter 11: Cartwheel Races 171

Chapter 12: Andrew Catch The Balloon 181

Chapter 1: Play Outside

The incredible out of doors offers youngsters and adults alike the possibility to have a super time while respiration deeply of sparkling air and soaking in all the blessings of the solar. Unfortunately, entire generations of children have moved maximum of their playtime indoors, focusing on laptop systems, television, and video games. Research suggests that the same old American infant spends most effective one half of of hour in outside play — a stark assessment to preceding generations who might in all likelihood frequently spend all day out of doors.

As a child, I used to like to play outdoor! I'd be up and out through 7:30 within the morning. Sometimes I'd live out all day, splendid going within the residence for meals. When a chum modified into with me, we might spend endless hours in out of doors enjoyment. If we bored with the video games we knew, we'd make up our non-public.

Even as soon as I become on my own, I may additionally need to discover something fun to do outdoor. In this economic catastrophe, you can discover some of the fantastic outside video video video games.

Outdoor play may be definitely as lots fun as gambling internal. It gives many advantages because of the bodily nature of most outside video video games. They can assemble cardiovascular energy and assist fend off weight problems. Spending time inside the exceptional out of doors will increase your solar publicity, it's far the excellent way to get Vitamin D, an critical detail to boost your mood and preserve off coronary heart sickness, diabetes and degenerative eye ailments. Outdoor play can help children increase gross motor talents, boom their balance, enhance their coordination, and assemble muscle power.

Time outdoor is soothing to frame, thoughts, and spirit. Research indicates that outdoor play can lessen the signs and signs of ADHD.

Schools with environmental-quality programs have shown enhancements in essential thinking and listening skills, and characteristic mentioned higher outcomes on standardized assessments. Research additionally indicates that playing freely in a green environment can lessen kids's stress, tension, and depression levels.

Nature is a amazing healer. When you are surrounded thru timber, glowing air, songbirds, vegetation, and different out of doors factors, you could discover you've got a extra effective mindset, a stronger imagination, closer relationships, and advanced social interactions. The Youtube video, Benefits of Playing Outdoors: Let Kids Get Down and Dirty thru manner of Kimberly Blaine describes extra benefits of playing outdoor.

Around the World

Number of gamers: 2 or greater

In this basketball recreation, honestly anybody takes turns taking pix baskets from positions that circle throughout the basketball hoop. To circulate from one aspect to the following, game enthusiasts ought to correctly make the shot in their contemporary function. When a player misses a shot, she or he gets one greater threat to make it. The capture is, If the person misses again, that participant need to bypass again to the primary spot and begin over. I had lots of amusing with this activity growing up.

Ball Race

Number of gamers: 2 or extra

For this amusing racing sport, you can want a massive, open discipline, at the least game enthusiasts, and one football ball or kickball consistent with player. The gamers start out collectively, difficulty with the useful resource of using side . The aim is to kick the ball to the surrender line without crossing some other player's ball. Whenever this takes area the participant have to start again. The first player to kick their ball all of the way beyond the end line is the winner.

Classic Outdoor Sports

Number of game enthusiasts: 2 or more

Some of the maximum commonplace outdoor sports sports are the traditional sports. Basketball, football, kickball, tennis, dodgeball, whiffle ball, badminton, soccer, volleyball and hockey can all be tailor-made to meet your degree of play and the form and period of your gambling area. Many shops promote tool that permits you to play multiple sports the use of the equal device, but you could additionally make do with what you have got available. These conventional sports activities activities are positive to entertain youngsters for hours on give up and might without trouble burn off all that excess strength.

Coin Toss

Number of game enthusiasts: 2 or more

This fun counting game requires a handful of change, a plastic container and a small kiddie pool. Prepare this recreation with the aid of the usage of filling a pool with water after which floating a plastic vicinity in the pool. Have all gamers stand a pair ft far from the pool and divvy out the trade evenly. The workout includes people taking turns tossing a coin into the pool, with factors presented once they sink a coin into the container. Players keep tune of their person scores. Pennies rely as one factor, nickels are 5 elements, and so forth. Whoever ratings the extraordinary amount of factors wins! You ought to make this sport more difficult, or make it easier by using using changing the diameter length of the container.

Flashlight Tag

Number of gamers: 2 or extra

This 1/three twist of tag is wonderful finished in the darkish. One participant is picked to be "it" and holds a flashlight in hand. The other game enthusiasts have to run faraway from the player who's "it." "It" tags exceptional gamers by means of shining the flashlight beam on them. The final player status will become "it" for the following round.

Football Fortune

Number of game enthusiasts: 3 or more

Want to sharpen your math abilties? This exercise requires a huge open area, a football, and at the least 3 gamers. It moreover lets in to have someone who's inclined to keep score. Prior to starting the sport, all game enthusiasts want to decide on a triumphing score price, and pick precise numbers on the way to motive game enthusiasts to lose all of their factors alongside the way. One participant throws the football and yells out a factor rate. The player who catches the ball wins that range of factors. The first character to acquire or exceed the prevailing range wins. Of route, if you entice the ball and your score provides as an lousy lot as a dropping amount, you lose all your elements and must start over again.

Four Square

Number of gamers: four or greater

Four rectangular requires no longer much less than 4 game enthusiasts, some chalk, a huge ball, and a decent sized playing region. One participant draws a huge rectangular after which attracts lines in the course of the center to make four smaller squares. Each player claims a rectangular. Players then take turns hitting the ball into each other's squares. Players are eliminated even as they are no longer capable of hit the ball out of their rectangular with their arms earlier than it bounces a second time. The winner is the remaining one popularity.

If there are extra than 4 gamers, people can rotate into the game whenever a participant is eliminated. The exercise maintains till all the sport enthusiasts have had a risk to play and certainly all and sundry however one has been eliminated.

Freeze Tag

Number of players: 2 or extra

In a few specific fun twist of tag, one player is picked to be "it" and starts offevolved out thru chasing the alternative gamers as they run away. When a participant is tagged, he or she ought to stand though in a "frozen" function. The great way a participant can can waft over again is to be tagged all over again with the aid of every other player who's on the run. The recreation is over while all gamers are frozen.

Group Tag

Number of game enthusiasts: 2 or more

In this twist on the traditional recreation of tag, one player is picked to be "it" and begins offevolved out with the beneficial aid of chasing the alternative game enthusiasts who run away. When a player is tagged, in desire to sitting out, she or he joins the "it" player to chase the others. Every special participant who is tagged does the equal until there is a large group of youngsters chasing just a few.

The endeavor is over at the same time as all and sundry is a part of the group. The final man or woman tagged can turn out to be the extremely-contemporary "it" or you may use some different criterion for this option.

Hide and Seek

Number of game enthusiasts: 2 or greater

Hide and searching for is a conventional interest that children and adults of each age can experience. This undertaking can be completed internal, however nowadays we'll interest at the outside range. Pick one participant to be "it" and characteristic that individual count as much as 30 even as the rest of the players run off and cover. At the give up of the rely, the person that is "it" is going on the hunt for actually each person who is hidden. Whoever is determined very last is the subsequent person to be "it." For a amusing model, strive gambling in the darkish. If you're gambling this recreation outdoors, it's miles surely beneficial to set stable barriers so game enthusiasts can't wander into chance inside the darkish.

H-O-R-S-E

Number of gamers: 2 or greater

If you have got were given a basketball hoop, H-O-R-S-E can offer you with hours of a laugh. One player starts offevolved offevolved out thru taking a shot. It may be an clean shot or it may be loopy hard. If the player makes the shot, the alternative gamers want to try the right equal shot from the equal vicinity. When a player misses, she or he receives a letter, starting with the "H" of "horse." When a participant reaches the "E" of "horse", he or she is out of the game. The remaining player reputation wins.

You can add interesting twists to this recreation, via capturing with one hand, fame on one leg, or via taking photographs collectively together with your returned to the basket.

Jaws

Number of game enthusiasts: three or greater

The game of "Jaws" calls for a pool, a floating raft, and at least 3 gamers. One player is selected to be the shark and the relaxation of the gamers climb on pinnacle of the floating raft. The shark tries to get the alternative gamers off the raft by using any way possible. Once a participant is knocked off the raft, she or he want to get out of the water. The very last player on the raft wins and gets to be the shark in the next round.

Marco Polo

Number of gamers: 2 or greater

If you've got got get proper of entry to to a pool, your youngsters will love this amusing activity. Pick one toddler to be "it." This participant will near his (or her) eyes and depend to ten. Then, with eyes nevertheless closed, that player yells out "Marco". The other game enthusiasts must respond with "Polo" on the same time as swimming away. The infant who is "Marco" should try to locate the alternative game enthusiasts with the beneficial aid of the sound of their voices and tag them. The final character to be caught can be the following "it." Feel free to possibility distinct word pairs for "Marco," and "Polo," to offer this game a memorable twist.

Pillowcase Race

Number of game enthusiasts: 2 or extra

Also fondly called "Potato Sack Racing," that could be a outstanding outdoor recreation for every youngsters and adults. All you need is a big grassy location and an empty pillowcase for each participant. Set a beginning line and a finish line. Have all game enthusiasts line up on the beginning line, repute inner their pillowcases and maintaining them up as immoderate as viable. When a pacesetter yells, "Go!" all the sport fanatics have to hop their way to the surrender line. This is a hilarious pastime for the observers, as they watch their buddies lose their stability and topple over each other. The first participant to make it over the forestall line in their pillowcase – and on their ft – is the winner.

Pool Categories

Number of gamers: 3 or greater

This amusing racing undertaking calls for a pool and at the least three gamers. The player who is "it" stands on one facet of a pool on the equal time due to the fact the final gamers stand on the alternative. The "it" participant makes a choice on a class, which incorporates colorings, cereals, films, and lots of others. Once the category is chosen, the "it" player is going underwater whilst the very last gamers pick out a phrase in that class and speak it aloud. The "it" participant then emerges from the water and starts offevolved to name things underneath their magnificence.

For example, if the magnificence is "colours," the player inside the water may respond thru saying, "Blue, inexperienced, yellow..." When gamers pay interest their selected colour they need to race to the opportunity side of the pool. The first participant to gain the opposite aspect of the pool and communicate to out the word they selected turns into the "it" participant for the subsequent round.

Twenty One Points

Number of gamers: 2 or extra

It's additionally fun to play basketball with the family or a collection of pals. You genuinely divide the institution into even groups and then play closer to every exclusive, abiding via the rules of a well-known basketball pastime. The first group to reach 21 factors is the winner.

Water Balloon Bombs

Number of game enthusiasts: 2 or extra

This undertaking is lots of amusing at an outside birthday party, BBQ, or own family reunion. It is maximum fun on a virtually warm summer season day, as long as you don't mind getting your garments moist. All you can need are a few filled water balloons. You can stash your prepared ammo in a kiddie pool or a massive cooler.

There are many approaches to play with water balloons. In addition to unstructured play in which you chase every special across the out of doors, lobbing balloons at all and sundry internal range, here is a sport that is easy, however a laugh.

Two game enthusiasts stand managing each one-of-a-kind, a pair yards apart. The beginning player selects a water balloon and tosses it to the alternative person, who catches it and throws it once more. After a pair of a success catches, the game enthusiasts take a step backward. The reason is to appearance how an extended manner apart they may stand to pass the balloon in advance than it breaks.

You can wreck massive organizations into corporations of players and undergo a device of elimination. The last corporation popularity the farthest from each different with out breaking any water balloons wins.

Water Gun Battle

Number of game enthusiasts: 2 or extra

If you discover your self on a heat summer season afternoon with no longer something masses to do, why now not load up some water weapons and begin a water gun battle? This can be an easy manner to get in a few exercise and cool down on the equal time. There's no longer whatever more fun than seeking to avoid getting moist on the identical time as looking for to spray water on everyone else! It's a amusing hobby that every one people of your family can experience.

Water Sports

Number of gamers: 2 or more

Almost any exercise this is finished on land may be done within the water, but basketball and volleyball are the most commonplace. To play the water version of a game, you may either invest a bit coins in specific tool, or you can use property you have already were given across the pool to your recreation. For water volleyball, you could keep close a regular volleyball net above the middle of the pool or just string a line at some stage within the middle. You need to make a basketball hoop out of a pool noodle, you may use a Nerf hoop and maintain close it over one end of the pool, or you may waft an internal tube as a basket. An internal tube makes a hard basket, because it floats spherical and doesn't live in a single location!

You can also use something water toys you already should create suitable device for other water video video games. Pool noodles and a whiffle ball or a beach ball make for exciting water baseball, with bases marked as spots on the perimeters of the pool.

Just final weekend, my family and I went swimming and used a seashore ball to play volleyball within the pool. We didn't have a actual volleyball internet so we positioned the extendable pool skimmer within the path of of the middle of the pool, laying it at some point of the seats of lawn chairs, one on every facet of the pool.

Water Tag

Number of game enthusiasts: 2 or more

Here is a amusing outdoor recreation if you have a massive outside and an prolonged hose with a spray attachment. You will want at least game enthusiasts for this sport, irrespective of the truth that the more people, the greater a laugh. One player will carry out the hose and try to get rid of every body via the usage of spraying water on the other gamers as they run away. The ultimate man or woman repute receives to feature the hose at some stage inside the next spherical.

Chapter 2: Rainy Day Indoor Games

Outdoor play is first-rate, however what approximately the ones days at the same time as the climate's nasty? What do you do at some stage in the prolonged, cold wintry weather months to burn off burdened power? Sure, you may play out of doors, however subsequently, you'll start to feel the bloodless.

What do you do whilst night time time time falls and your children but have lots of power? This is in which indoor play can are available in particular reachable! Although it can't be as physical stimulating as out of doors play, indoor play although has its rate. It can help youngsters boom their social and emotional abilties, and may enhance their creativeness. It also can though rely as exercise, because of the reality many indoor video games even though require a few motion. In this financial damage, you may find out the exceptional indoor games that your little one can play with distinct children or with adults.

Balloon Slap

This is a slow-motion model of volleyball that even more youthful kids can participate in. To play this exercising, you can want at the least game enthusiasts, some string and a balloon. Pick a place in your house wherein there may be hundreds of open area and put off any breakables. Attach the string from one wall to the opportunity with the resource of taping the ends to the wall with protecting tape or painter's tape. Fill a balloon with air and location as a minimum one player on every side of the string. The motive of balloon slap is to have every participant serve the balloon backward and forward over the string without letting it hit the ground. The extra gamers you have got, the more balloons you could upload to boom the project.

Board Game Night

Classic board video video games the form of Life, Sorry, Monopoly, Candy Land, and so forth. Never get antique! Set aside a night time as family board recreation night time, in which in truth all people gets to play with every tremendous. A in reality cool idea is to offer snacks which might be associated with some issue recreation you're playing. For example, if you're gambling Candy Land you may offer snacks like jolly ranchers, rock candy, ice cream sandwiches, chocolate pudding, and so on.

You may additionally have friends over for game night time and make quite a few video video games available for them to play in precise additives of the house. With snacks and some component to drink, everybody can mingle, play the video video games that interest them, and commonly have an fantastic time.

Bouncy Ball Toss

Bouncy Ball Toss is a shape of hybrid, a mixture of darts and interior basketball however a whole lot extra hard and interesting than without delay basketball or darts with the resource of themselves. To play Bouncy Ball Toss, you could want some brown paper baggage or empty packing containers, some small bouncy balls and a black marker. If you're the usage of paper baggage, you'll moreover need some thing to crush every bag.

To put together for the game, use the black marker to install writing point values on each bag or container. You can then set up the boxes in any configuration you'd like. I in my view determine on a pyramid form because it makes the sport the maximum tough, but you may set up them in a line, in a geometrical shape, or any way you pick. The intention is for every player to bop the ball on the floor in this type of way that it lands in one of the boxes

To make the game even more tough, you can attempt playing on one-of-a-type surfaces. Bouncy balls have a tendency to dance greater immoderate on solid surfaces, that could deliver them a interesting speed. Carpets will be inclined to provide greater of an aiming assignment because of the fact the balls obtained't soar as immoderate and often will deal with interesting random tangents. You also can try gambling on an choppy tile or gravel ground to appearance what takes vicinity.

You also can play this challenge on a flight of stairs if you don't have containers to be had. In this method, game enthusiasts take turns throwing the bouncy ball in the direction of the better stair steps and counting the huge style of instances it bounces earlier than it hits the ground. I become an avid collector of bouncy balls when I became a toddler and often carried out this recreation on the stairs. Since then, I've taught the steps model to my more younger cousins who're five and three and that they adore it.

Creative Drawing

For this recreation, all you want is paper to attract on and some thing you may draw with, from crayons to markers to pencils. In this recreation, one participant starts offevolved offevolved with the resource of way of drawing some aspect on the paper and passing it to the subsequent character who can add to it. This activity is quite amusing in huge agencies as it's exciting to look what particular gamers have brought to the drawing by the time the paper gets decrease returned to you.

To keep absolutely everybody occupied, deliver all and sundry a sheet of paper, so more than one drawings is probably surpassed during the business company, simultaneously. People can proportion colors, promoting cooperation, or be assigned a unmarried coloration for the length.

Family Bingo

This fun spin on bingo can hone your toddler's memory skills and is right for greater more youthful children who are nevertheless mastering about their family members. All you could want is 9 pix containing pix of your own family participants and 9 small devices that could characteristic markers, along with buttons. Start by using using assembling the snap shots in a 3x3 rectangular. Instead of calling out numbers, you name out names which consist of "Mom," or "Grandpa," and if the child has a picture of the member of the family you name, he or she places a marker on pinnacle until a line is created. You can play this undertaking with any amount of youngsters through honestly developing the amount of pictures you use with the beneficial useful resource of increments of nine. Playing with a couple of players is amusing due to the reality not everyone can also have the identical pix. This mission is also amusing for an older sibling to play alongside with his or her greater younger siblings.

Floor Tile Checkers

Everyone is familiar with the time-examined game of checkers, but who says you need to go out and buy the valid board game? If you have were given a floor in your own home that has a sample of squares, you could with out troubles turn it right right into a massive checkerboard! All you could want is a few protecting tape and a few aspect that could represent game portions, together with small balls or rocks. This undertaking is certain to entertain little kids on a wet day!

Use shielding tape to mark off a place of the floor a good way to constitute the board. You can then area an opening of tape through the center of the "board" to designate each player's side. Using toys as distinct game portions, every infant attempts to get their portions to the alternative participant's issue as in a traditional checkers workout. You can alter the scale of the "board" to fit your selections, or to accommodate the dimensions of the room. For younger youngsters, you can start off with a smaller board to simplify the sport while they examine it.

Go Fish

For this undertaking, all you want is a deck of gambling cards. Shuffle the deck and characteristic one family member pass out seven gambling cards to each participant.

Place the last playing playing cards inside the middle. Everyone seems at their arms and pulls out any matching fits (the identical card rate; sevens, kings, and so on.) to set on the desk. One man or woman starts offevolved offevolved the game via asking if every person has a outstanding card. If no individual has that card fee, the asker has to "skip fish" via taking a card from the pinnacle of the deck, then the following player takes a flip. However many humans have that card rate, they're obliged to provide it as an awful lot because the participant who requested, and that man or woman gets to invite for some different card. The first man or woman to finish their hand thru matching suits to all playing playing playing cards wins.

Homemade Connect the Dots

This recreation is right for toddlers and small kids who are notwithstanding the truth that developing their first-rate motor abilties. It's a brilliant manner for mother and father and their children to bond and function fun together. For this pastime, all you will want is multiple large poster board sheets, a fixed of markers and some colorful circle-normal stickers. You should make patterns on the poster board by means of the usage of the use of setting the stickers strategically and the goal is in your infant to use the marker to connect them. As your infant grows or starts offevolved offevolved to understand fundamental connecting abilties, you can make the purpose more hard with the beneficial resource of the use of particular colored circle stickers after which encouraging your baby to attach best the same-coloured circles to assemble patterns which are extra complicated. It can also be a laugh to have your infant attempt to bet what the final give up end result may be based mostly on wherein you've located the stickers.

Indoors Bowling

This a laugh pastime is fine for wet days and is a price-effective way to have some amusing.

Line up some empty water bottles in an open vicinity, and then find out a medium-sized, plastic ball. Take turns rolling the ball to knock down the water bottles as if they may be bowling pins. If you want a terrible line, lay down a few covering tape or blue painter's tape; it's miles actually clean to pull this up off the floor whilst you are finished.

You can range the trouble with the aid of changing the dimensions of the ball, through playing on carpet in place of hardwood, of via manner of the usage of filling the bottles partway to guide them to more difficult to knock over.

Indoor Hoops

This exercise is a model of basketball, best it's executed interior. The best system you want is a few issue to feature a ball and some thing you could use as a hoop. You should purchase small balls and hoops meant for bed room play, but I pretty propose being innovative and the usage of some difficulty substances you have were given handy.

You can steal the place of job version of this sport for home use. It consists of crumpled up balls of paper and a wastebasket because the hoop.

You can also use this sport to inspire your youngsters to preserve their room clean with the useful resource of the usage of grimy garments because the ball and a laundry basket due to the fact the ring. To inspire your youngsters to throw away their rubbish, you may allow them to toss their empty cups, drink boxes, plates, and plenty of others. Into the rubbish can.

Jigsaw Puzzles

Jigsaw puzzles are usually inexpensive, they could pose a superb task, and might provide hours of exciting amusement and suitable communication. Puzzles are exquisite for playing on my own or with a few distinctive pals. Jigsaw puzzles are available many designs. They range from the easy to the pretty complicated. I've even seen a 3-D jigsaw puzzle that permits you to acquire a castle.

Jigsaw puzzles can regularly be academic. One of the extra famous designs is a map of america this is illustrated to represent what is specific approximately each state.

When kids artwork on a jigsaw puzzle collectively, it encourages teamwork. Puzzles can become extra interesting for small children, the closer they arrive to of completion.

Name The Song

To play this pastime you can need some recorded song. This pastime is a fantastic manner to get children engaged in the global of music and share your preferred songs with them. One player, normally the decide, plays just a few seconds of a song. The purpose is for the alternative participant, normally the kid (even though it is able to be fun for the child and discern to switch roles), to wager what the song is, primarily based absolutely totally on simplest listening to some notes. This sport also can be played amongst kids. Whoever can bet the maximum songs efficiently wins.

This may be a fun birthday celebration assignment in which anyone is asked to hold their favored song and a way to play it. Players can take turns playing brief clips of their preferred songs, developing the clip period until a person guesses the track.

Old Maid

For this game, you may want a deck of gambling gambling playing cards with three queens taken out. The final queen is the "antique maid" card. Once there is simplest one queen left in the deck, one character will deal out the whole deck gently to the opposite game enthusiasts. After genuinely every body has their playing cards, they undergo their arms and lay down any matching suits.

The reason of the game is to make the most fits with out being stuck with the "vintage maid" at the save you of the hand. The game enthusiasts take turns, starting with the participant to the left of the dealer, picking one card from the player to their left and putting apart their non-public matched fits. The hand ends while one participant has no extra playing playing playing cards.

The man or woman left preserving the "antique maid" mechanically loses the hand. The sport can be accomplished handiest for the fun of seeing who sooner or later ends up with the "vintage maid", or you could maintain rating across severa hands based totally absolutely on the variety of fits each participant has, excepting, of route, the player defensive the "vintage maid" card.

Restaurant Memory

This is a a laugh undertaking to play with the children at the same time as you're out to eat at a eating place. Most ingesting locations provide youngsters under 12 with a cool lively film-like menu or a placemat. After setting your orders, permit your baby to look at the photos at the menu and ask them to memorize what they could. After a few minutes, take the menu, hold it in the the front of you and take a look at your toddler(ren)'s memory via asking questions related to the menu which incorporates, "How many tentacles does the octopus have?" or "How many bubbles is the cat blowing?" If you're playing with a couple of toddler, you can maintain track of how many of factors everybody has and make the winner the primary character to acquire 20 factors…or you can see who has the most factors while the food arrives.

Silent Ball

This is a amazing sport to play whilst dad and mom need a few peace and quiet. The gamers sit scattered round a room. The pointers are that the gamers need to skip a balled-up clean sock to each different but no individual should make a valid. If absolutely everyone makes noise, that man or woman is out. Additionally, if a player can't capture the ball, that participant is out, too. The workout will become the most amusing even as it gets all the way down to the ultimate gamers who are competing tough to become the winner.

Spoons

Spoons may be a whole lot of amusing and can result in a hilarious free-for-all. For this sport you'll want a deck of playing cards and one fewer spoons than you have got game enthusiasts. Each participant is dealt four playing playing cards and the the relaxation of the deck sits earlier than the provider. Place all the spoons (e.G., when you have 5 game enthusiasts, you may place 4 spoons) within the center, with handles pointing outward. To begin the sport, the supplier alternatives up one card from the deck, chooses one of the now 5 gambling playing cards within the hand to discard, and passes the discard, face proper all the way down to the participant on his or her left. This player then selects a card to discard, handing it to the participant on the left. The final participant earlier than the enterprise, locations his or her discard proper right right into a separate trash pile. The company maintains to pick out up a card and discard one to the left till the hand ends.

If you, as a participant have accrued four of the identical card, it is time to choose up a spoon. You can be sneaky and slip one far from the affiliation with no person being the wiser, or you could openly draw close one within the open. At any price, you preserve on passing playing cards spherical.

As fast as you, as a participant, have a look at a spoon is missing, you can take maintain of a spoon. What regularly effects is a grand melee in which wrestling suits turn up, palms are scratched, and spoons are bent. Ultimately, one character is left without a spoon, which means they get a letter, starting with "S" and persevering with on through "P", "O", "O", and "N". Once a player loses 5 hands, they'll be out of the sport, along with one of the spoons. The playing cards are dealt another time, the spoons are returned to the center, and play resumes. The recreation keeps till, of the final gamers, one is eliminated.

Stair Ball

Number of gamers: 1 or greater

While we're talking about stairs, that is a top notch way to play solo and feature some fun in the method. Simply take a ball and bounce it toward the better stair steps. The goal is to keep the ball in play as long as feasible with the aid of the usage of hitting it once more up the stairs even because it comes toward you. The ball remains in play till it hits the floor.

You can play this with friends by using timing the amount of time the ball stays in play or through counting the amount of hits earlier than the ball hits the ground. The winner is the simplest with the longest gambling time or the handiest with the most hits, relying on what regulations you are the use of.

Sticky Ball

This fun indoor recreation is thrilling for more younger youngsters and promotes coordination abilities. All you want is a roll of painter's tape or protecting tape and paper which you don't thoughts crumpling up into balls (newspaper works brilliant for this). Choose a extensive doorway in your own home and region portions of tape, sticky factor inside the path of the gamers. You can area it in immediately strains or create a web-like sample. Then each participant takes a flip throwing crumpled balls of paper on the tape in hopes that they may stick. Each stuck ball earns each participant one issue. The participant with the most factors after each person has thrown their balls wins.

Thumb Wars

This has been a well-known project among children for lots generations. It is a time-examined traditional that promotes hand-eye coordination without the use of a tv or a enterprise controller. Thumb wars starts offevolved with game enthusiasts, despite the reality that youngsters regularly have tournaments to determine the champion of the organization. Two gamers begin thru manner of sitting, or fame, head to head and hooking their palms across the opportunity person's arms in a corporation grip that leaves the 2 thumbs sticking up. The reason of Thumb Wars is to use your personal thumb to pin down your opponent's thumb. Once you've have been given your opponent's thumb to your preserve near, you need to maintain it down at the same time as you rely out loud to 3 before the victory turns into yours.

Chapter 3: Group Games

Team-sized video games can be the maximum amusing of they all! Many of those video video games may be tailored for inner play, which gives them flexibility for all sorts of sports, consisting of birthday occasions, circle of relatives reunions, university fairs, and potlucks. You can use them for a "non-event" too, along with even as you find out your home complete of youngsters and need to come up with a few component to do, and rapid! The video games I've blanketed in this chapter may be cherished via youngsters of every age and are brilliant for instances while you want to add a hint pizzazz for your lifestyles.

Because those video video games are achieved in companies, they gift an great possibility to make bigger vital social capabilities and teamwork. Team-constructing video games help decorate one's morale, extend latent control, take a look at with creativity, and enlarge together in the capability to successfully remedy issues.

However, that's no reason to expect that they fall quick inside the amusing class. As I stated, company games may be the maximum fun of them all!

Charades

In this fun performing sport, one infant alternatives a phrase or phrase and then acts it out with gestures simply so the opportunity children can art work to wager it. No verbal cues are allowed in this enterprise; only frame language, facial expressions, and gestures. The character who guesses the right word or phrase gets to act out the following one. This is a a laugh game for adults as well as mixed-age organizations.

Cheese Puff Blow

To play Cheese Puff Blow, you will want covering tape, a bag of cheese puffs, one straw for each participant, and sufficient game enthusiasts to break up into at the least teams. If you want to characteristic up factors, you can place a hand-drawn goal categorised with element values, at the stop line. You'll moreover need to clean a desk for this recreation or, I assume, you may play on a floor space.

The game enthusiasts ought to interrupt into organizations of human beings each. You can use covering tape to mark the begin and stop lines at contrary ends of the table. Beyond the give up line, you can tape down the motive. Each group options one player to face off within the route of the other group. Both gamers should set a cheese puff within the lower back of the start line. When an appointed player says, "Go!" every gamers have to blow through their straws to move their cheese puff toward the give up line. When the primary cheese puff rolls throughout the end line, each gamers save you blowing and permit it roll onto the purpose itself. The first group to earn 20 factors is the winner.

Dice Roll

To play this simple but amusing recreation, all you'll want is a die or pair of cube and one piece of paper for each player. Each participant takes a flip rolling the dice and records every quantity they roll. The item of the game is to feature the numbers together and be the primary person to achieve one hundred. As gamers approach one hundred, the aggressive side every so often surely comes out!

Draw and Fold

Draw and fold requires no less than 4 game enthusiasts, a pen or pencil for everybody, and one piece of paper for each spherical. It is first rate if all of us sits in a circle. To start, the number one player writes a sentence on the top of the paper after which passes it to the participant on their right. The 2nd participant then rewrites the sentence but within the form of drawings. For example, if player number one wrote "I love you," then participant range ought to draw stick figures with a heart in amongst them. The second participant then folds the pinnacle of the paper so that the written sentence is hidden and best their example is visible earlier than passing it to the 1/3 participant. The 0.33 player then should attempt to decipher the drawing thru using translating it right right into a written sentence and folding the paper to cover the drawing. Player 4 then need to try and translate the cutting-edge sentence into a current drawing. Players preserve passing the paper round till each the paper is used up or the circuit is completed. At the prevent, anyone could have a have a look at the whole piece of paper to reveal all the answers. Like Telephone, the final consequences can frequently be hilarious.

Family Jeopardy

Does your circle of relatives love trivia questions? If so then this can be a tremendous recreation to play at circle of relatives sports. To prepare for this recreation, you can write up questions and solutions about your circle of relatives in advance. Questions can also consist of matters inclusive of, "What one year did Sally graduate from college?" or "What is Dad's middle call?" You can spoil the questions into categories as they do inside the actual Jeopardy show, and categorize them via family member or with the aid of problem. You want to moreover assign a difficulty price to each query. Depending on the scale of your family and the age of the people, each person can play in my view or work as a part of a group. After all the questions are exhausted, the individual or group with the maximum factors wins.

Freeze Dance

This delightful game can be finished solo or with a group of friends. Pick a few fun song. The rule is that everyone should forestall dancing at the same time as the music is became off, no matter what role they may be in. If a player moves or falls out of role, he is out.

Design a playlist that includes the most well-known kids' songs. If you have got got a few lively youngsters, this is a splendid manner to put on them down. Freeze dance also can be accomplished the usage of chairs scattered around the room(one fewer than the huge shape of members), as a fun tweak to musical chairs.

When the song stops, anybody scrambles for a seat and the individual left with out one is eliminated. Before each spherical you may take a chair away till there is notable one left, and whoever gets that last seat even as the tune stops is the winner.

Guess the Drawing

This mission calls for big portions of paper and a few markers. Hang up smooth sheets of white paper on the wall. One participant goes as a whole lot because the paper and starts offevolved offevolved to draw some factor while the relaxation of the gamers yell out guesses as to what the drawing is. Whoever can bet the right phrase first receives to be the subsequent man or woman to attract a phrase.

This can be without difficulty became a group recreation with the resource of using giving the artist for every group the same phrase. The institution that guesses the word first wins a factor for the team and every other artist for each organization is chosen for the subsequent spherical.

Hot Potato

For this activity, put on some a laugh tune and pass round a moderate item, like a beanbag or a small ball. Whoever has the object in hand while the music is grew to turn out to be off is "out." When it's down to 2 humans, the person without the "heat potato" whilst the tune ends is the winner..

Human Twist Puzzle

In this fun teambuilding pastime, all contributors stand in a circle. The great duration for this interest is about nine human beings, so when you have a large bunch of folks that want to play, you can want to divide them into smaller groups.

At any rate, train the gamers to walk forward and hold close to the right hand of the man or woman within the opposite component of the circle. When everyone has done this step, ask them to clasp the left hand of a person close to them.

This will bring about all people associated in a tousled mass. The purpose of this enterprise is to art work together to untangle the mess with out letting skip of anybody's hand. When solved, the people want to be all yet again in a circle, defensive arms.

As you can don't forget, this recreation requires some near bodily touch. Depending at the age of the human beings, some humans can be maximum comfortable in a equal-intercourse group, and others may not be snug with any bodily touch in any respect, so live touchy to the dreams of your site visitors. Generally, in case your individuals may be snug playing the game "Twister" they'll don't have any problem with this hobby.

I Would Rather

This recreation is a first-rate icebreaker, as it allows young youngsters discover others with whom they have got subjects in commonplace. All you need for this sport is a list of questions. Each query must require the players to pick among two options and to show their desire thru moving to a designated factor of the room. All gamers start through strolling to the center of the room. For example, if the query is, "Would you rather placed on a baseball cap or a visor?" you will designate one aspect of the room for visors and the other facet for folks who select out baseball caps. This can preserve with a few more questions that are home made to help children studies more approximately each one-of-a-kind.

Improv Skits

This pastime is notable for a huge agency of humans and it's far going over well with blended-age companies. Preparation within reason easy; truely fill a few paper luggage with random gadgets from your house. Include a aggregate ranging from articles of apparel to kitchen utensils to something in your rec room that you don't thoughts people handling. Divide the contributors into organizations of round six human beings and supply each group a bag of random gadgets. Give each institution a few minutes to design a brief skit the utilization of every of the objects within the bag; then come back collectively to have a look at every unique's skits. Hint: the greater random the objects in a bag, the funnier each skit is sure to be.

Mummy Wrapping Contest

This a laugh game requires at the least gamers who are inclined to be mummies and at the least 4 game enthusiasts who're willing to break into groups and race to wrap their mummy. The exceptional device required is a roll of rest room paper for every group. Each mummy stands though on the same time as the teams use a whole roll of relaxation room paper in a race to wrap their mummy the fastest. Keep a virtual digicam handy to snap some funny photos.

Simon Says

"Simon Says" is each other traditional sport that children love. Pick one little one to be "Simon." That toddler will then call out commands, maximum of the time beginning with the phrases, "Simon says." For instance, one command is probably "Simon says contact your nose." If "Simon" does now not begin the command with those key terms and a participant plays the command except, she or he is out.

One variation of this game this is loved with the resource of greater younger youngsters is to replace the name "Simon," with the decision of the person that is in fee of giving the commands. Kids love the idea and it's far powerful to generate some laughs and giggles from the agency.

Stick Dance

This undertaking calls for an ordinary fashion of game enthusiasts, some tune and an prolonged stick, a brush, or a yardstick. Each participant selections a dancing accomplice. The extraordinary man out (or female) dances with the stick. Somebody no longer taking element in the game may be in rate of turning the music on and off. The song participant begins offevolved offevolved the song and we should all people dance with their associate for some seconds before preventing it. When the song stops, anybody ought to forestall dancing and switch companions. The participant who's left with out a companion ought to then dance with the stick until the music stops another time.

Team Story Telling

The thing of this hobby is to gather all the humans collectively and allow each participant take a flip at adding to a tale line. For example, the number one participant will start off with the useful resource of growing the first sentence – or phrase – of a tale. It can absolutely be something. The subsequent participant will then upload a second sentence or phrase to the story. Go throughout the group, letting each person contribute a factor, till the tale is wrapped up by manner of the very last participant.

The mission can come to be unwieldy and those can also moreover become bored if there are too many human beings. If you have got greater than a dozen human beings, I advise you chop up up into separate groups and allow each institution broaden a separate story. The tale can be carried out out loud or it can be written down a sentence at a time on a piece of paper and handed from man or woman to individual, to be read out loud at the give up. Players can be exceedingly progressive, leaving the tale putting after introducing an unusual word.

To upload to the volume of problem, have every player write the decision of someone, a place, or an item on separate quantities of paper. Then combo up the paper and function everybody draw an object to apply as part of their contribution to the tale.

Telephone

This fun task is proper for while children are collectively in corporations of at least 5. The first little one or adult thinks of a word and then whispers it into the subsequent participant's ear. That player then whispers it into the following participant's ear. The final player speaks the phrase aloud to peer if it matches the precise phrase. Sometimes the phrase gets so mashed up that the give up quit result is hilarious. This changed into every different amusing activity that I finished as a infant, and a few times I took excellent pride in changing the phrase purposely causing the entire elegance to giggle on the quit.

The "Egg-cellent" Race

This high-quality birthday party pastime is super for activities with large agencies of human beings. All this is required is enough spoons for every player and one egg in step with organization. If you're gambling outdoor, it's a number of fun to play with raw eggs. However, if you're gambling indoors, I advocate you boil the eggs; otherwise it has the potential to get messy. Begin with the beneficial resource of dividing the organization into teams.

This is a relay race, so half of the organization ought to be on one facet of the sector and half of at the opportunity side. The two elements have to be about ten toes apart. The first person to play starts offevolved offevolved the race with the deal with in their spoon of their mouth and an egg inside the spoon. The race begins with that player balancing the egg at the same time as racing at some point of to the opportunity issue and transferring the egg into the subsequent participant's spoon. If the egg drops or breaks, then the group is removed. The first organization to finish all legs of the race with egg intact is the winner.

Treasure Hunt

What can be extra amusing and hard than looking for "buried treasure?" For this exercise, hide an object or a difficult and rapid of items someplace throughout the residence and leave clues. You can start off with the beneficial resource of leaving the clues in apparent places after which have every clue lead your child to the following hidden spot. You also can draw a map and have your toddler find out the treasure that way. This sport additionally may be a laugh on the equal time as finished outside. If the business enterprise is huge, you can want to organisation the game lovers into businesses, giving every team a one-of-a-type set of clues, or as a minimum scramble the order of the clues.

The Human Christmas Tree

This is a first-rate family activity to play around Christmas or Thanksgiving time and is confident to cause some laughs. Keep a virtual camera geared up to take photos.

For this exercise, all you can need are a few Christmas-themed arts and crafts materials, which incorporates leftover wrapping paper, stickers, tissue paper, or ribbons. One man or woman volunteers to serve as the "tree" at the identical time as the alternative gamers take turns the use of the cloth to "get dressed" the tree. If you're throwing a celebration, you could play the sport the use of party substances to "wrap" a "gift". There is not any aggressive objective to this exercise, so it is able to generate a few silliness and create splendid reminiscences for years yet to come.

Chapter 4: Solo Games

Sometimes there's surely nobody spherical to play video video video games with. This economic disaster is devoted to video video games that a toddler – or an man or woman, for that depend - can play by myself. Parents or guardians can also take part in the ones solo video video games, it really is great for building determine-child relationships. My Dad and I used to spend many weekends together, absolutely the 2 folks, but we constantly determined a few issue to do together for entertainment. This bankruptcy offers some exceptional video games that your toddler can play on my own or with you.

Bubble Gum Challenge

This exercising can be finished with one toddler or as a race among a couple of kids. All you want is a few wrapped bubble gum, ideally a hard-to-bite variety, and a couple of oven mitts, socks or irrespective of the children can in shape over their fingers. The Bubble Gum Challenge is to appearance how lengthy it takes to unwrap the piece of gum, chunk it, and blow a bubble.

Guess the Hand

This conventional and simple game is famous among small kids and might assist them develop the potential to recognize visual cues. Player one is the discern and participant is the child. To play guess the hand, all you need do is cowl a small object for your hand by way of manner of balling your fist, then ask your baby to bet which hand is hiding the item.

Home-Made Skee Ball

Who doesn't love the conventional pastime of skee ball? Instead of packing the circle of relatives up for a journey to the ever-so-pricey arcade, right here's the way to easily make a free domestic model the use of best a cardboard container and an empty egg field:

Your discipline ought to ideally be square and characteristic 4 flaps. Using a safety knife, remove the 4 flaps and then reduce the top of the sphere at an upwards mind-set. Tape the

4 flaps collectively after which attach it to the bottom perspective of the arena the usage of a few tape. Place an empty egg carton internal of the field to characteristic the ball catcher. It may be beneficial to connect the egg carton to the floor of the field so it obtained't move spherical or wander away. Using a few small, light-weight balls (bouncy balls do the trick and also can be used for a number of the opposite games in this ebook), each player takes turns tossing a ball up the ramp with the reason of getting all of their balls into the egg carton!

Hopscotch

Hopscotch is a high-quality solo undertaking and also can be completed outdoors to promote bodily hobby. It may be accomplished thru kids as greater younger as 4 or five. The first step is to attract a grid. If you're gambling outdoor, you could draw squares at the sidewalk with a piece of chalk. If you're playing interior, you can use covering tape to mark out the squares on the ground.

The most common hopscotch pattern is a unmarried rectangular, observed with the aid of two squares component thru detail, then one rectangular located via squares, however you could make it however you'd like. One alternative is to region honestly one among a type numbers indoors every square.

The player starts offevolved by means of using throwing a small object along with a rock into the primary square. The rock need to land definitely internal. The player then starts offevolved offevolved to hop thru the path, landing on one foot for single squares and toes for double squares. Once the participant reaches the cease, she or he turns and hops lower back, grabbing the rock alongside the way. Then he or she tosses the rock into the following square and repeats the sample. If the player steps on one of the traces or outdoor of the squares, she or he need to start over. To make this recreation extra hard, you may upload variations to the guidelines. For instance, you may set a rule to pass over the rectangular maintaining the marker.

How Fast Can You Go?

An best little one can flip almost any regular mission right right into a fun recreation with the aid of turning it right right into a timed project. Challenge your infant to look how prolonged it takes him or her to dress, to scrub the dishes, to take out the rubbish, to pick out up their room or every other venture that would in any other case be regarded as a monotonous chore. Turning chores into video video games can assist your infant advantage aggressive abilties and inspire her or him to attempt for development.

Ice Digger

The aim of ice digger is to free a small treasure from a small block of ice using only the materials supplied. To put together for this pastime, take a disposable cup and region a small non-floating object in it earlier than filling the cup with water. Once the water is absolutely frozen, you can then come out the ice or peel the disposable cup away to show the block of ice. Provide your infant with device together with a small quantity of heat water, a spoon to chip away the ice or some aspect you might imagine of to help them unbury the frozen treasure. Unless you're gambling this outside, it may be useful to permit them to paintings on their ice over a shallow pan so the water doesn't get all around the location.

Magic Cup Game

You've probably visible this age-antique activity earlier than. The top notch detail about it's far that the quantity of game enthusiasts who can participate is flexible. An character can play it with one to a few small children, as plenty as four small kids can play it among themselves, or a toddler can venture him or herself by the use of the usage of playing it by myself. Take 3 plastic cups and positioned a small object below one, ideally some thing like a coin or a small stone. Then shuffle the cups spherical and have the participants wager which one includes the object.

If you are gambling with a set of kids, you use sleight of hand to exchange gadgets on the same time as a infant starts offevolved to capture on. You also can help them take a look at the sport so that you can trick their buddies. Almost any magic ebook let you research this trick.

Make Your Own Puzzle

Here is every other without a doubt amusing and time-examined enterprise. Let your baby draw a image after which lessen it up into puzzle quantities that may be with out problems reassembled. You can lessen it into easy squares and triangles or you can get in reality hard and reduce it into loopy shapes. You can also cut up images from magazines or newspapers to make an exciting puzzle. My dad used to make up puzzles for me as soon as I become little and I have become often entertained for hours via them.

Solitaire

Solitaire is a mythical unmarried-player card undertaking that can be executed with a deck of playing cards and is available as a unfastened activity on maximum laptop systems. To start, shuffle a large deck of playing playing cards and divide it into seven stacks. Set the stacks face down from left to right, then start to reveal playing cards face up as follows. The first row need to nice include a unmarried card, face up. The 2nd row ought to consist of playing cards, one face up one face down, the 0.33 row ought to have three gambling playing cards, one face up and face down, and so on. The seventh row must contain seven gambling playing cards inside the same fashion. Place the leftover cards to the element to draw from. The aim of Solitaire is to assemble each row from Ace to King in the same in form with each card alternating in shade. The YouTube video Solitaire Games: Solitaire Card Game Rules with the beneficial aid of eHowSports offers additional seen belief at the manner to play this endeavor.

Solve a Rubik's Cube

A Rubik's Cube is a small, adjustable cube that has a in reality one in every of a type shade on each of its six faces. Each face consists of smaller squares that can be moved through the use of turning the rims of the dice. The object of this recreation is first of all the colours scrambled after which go back it to sturdy shades on each aspect of the cube. This is the most effective endeavor I've protected that is critical to shop for, but it's miles extensively to be had in any maintain that sells video video games or toys.

The Box Game

For this undertaking, you could want a small lightweight cardboard field. Empty macaroni packing containers artwork well. The goal of the game is to pick out out the sector off the ground using your tooth and with out shielding onto a few aspect for guide. You can play this recreation on my own or you may embody the entire own family. It's fun and fun to appearance who can pick the sector up maximum efficaciously. If this sport is truely too smooth, you could create variations, consisting of choosing it up collectively collectively together with your tooth on the same time as popularity on one leg.

The Sensory Game

This sport is geared for small kids and is often excellent completed with a decide. You will need a bag packed with random devices from throughout the house. The toddler, with closed eyes, reaches into the bag, grabs an item after which tries to guess what it's far with the resource of using the opportunity senses. Older youngsters may be uninterested in this endeavor but it can be interesting for little kids.

Towering

Building a tower out of random devices can be amusing and an interesting mission, every for you and for any lookers-on. You can play in your personal or with buddies. I once in a while, even now, can be caught constructing a tower out of the items on my desk, while I assume my food at a eating place.

You can use gadgets with lots of weights, sizes, shapes, and textures and notice how excessive you could gather your tower. Small kids often can be entertained for hours with such things as paper cups, building blocks, cardboard packing containers or books. You can use truely any solid substance as a part of your tower.

Playing playing cards are via a long way the most hard fabric to apply because of the reality they may be so slight and smooth. When I come to be little, I used to stack the little character creamers boxes and be aware how excessive I may want to circulate in advance than it toppled over.

If you make a decision to play a towering sport with a couple of people, players can cut up into agencies or play for my part.

Whoever's tower is the super or stands the longest is the winner.

Wall Ball

If you want to exercising your catching abilties, wall ball is a amusing sport that you can play solo, or with others. All you need is a wall and a ball that bounces. Tennis balls usually art work well. Simply throw the ball in competition to the wall after which capture it even as it bounces decrease decrease returned. Throwing the ball tough can create an interesting mission.

Chapter 5: Educational Games

Who says all video video video games are just for a laugh? Or that reading can't be amusing? Games can provide an effective manner for every body, from the youngest kids to oldest adults to sharpen important bodily and intellectual abilities so that you can assist them every in college and in terms of "actual" lifestyles. Almost all the video video games we've already noted require a few shape of studying or talent improvement. Here's a precis of some of the benefits you can gain from the video games we've already talked about:

Developmental Focus

Game Title	Chapter	Hand-eye Coord.	Memory	Balance	Social	Problem Solving
Ball Race	1					
Balloon Slap	2					
Basketball Games	1					
Board Games	2					
Bouncy Ball Toss	2					
Box Game	4					

Bubble Gum Challenge   4

Card Games   2

Charades   3

Cheese-Puff Blow   three

Coin Toss   1

Connect the Dots   2

Creative Drawing   2

Dice Roll   3

Draw and Fold   three

Egg-cellent Race   three

Family Bingo 2

Family Jeopardy   three

Floor Tile Checkers 2

Football Fortune   1

Four Square 1

Freeze Dance   3

Guess the Drawing 3

Guess the Hand   4

Hide and Seek   1

Hopscotch   4

Hot Potato   3

How Fast?   four

Human Christmas Tree   three

Human Twist   3

I Would Rather   3

Ice Digger   four

Indoor Hoops   2

Indoor Bowling   2

Jaws   1

Magic Cup   4

Marco Polo   1

Mummy Wrapping   three

Name the Song   2

Outdoor Sports   1

Pillow Case Race   1

Pool Categories   1

Puzzles   2

Sensory Game   four

Silent Ball   2

Simon Says   three

Skee Ball   four

Skits   three

Solitaire   4

Stair Ball   2

Stick Dance 3

Sticky Ball 2

Story Telling 3

Tag   1

Telephone   3

Thumb Wars 2

Tic Tac Toe   2

Towering   four

Treasure Hunt   three

Umbrella Ball   three

Wall Ball   four

Water Balloon Bombs   1

Water Gun Battle   1

Water Tag   1

Games can help kids – and adults, too – growth their talents in spelling, vocabulary, sentence formation, math, geography, motor capabilities and social interaction. Most children already look at via way of manner of searching instructional television shows and playing video video games. Sometimes a clean sport can be used to educate kids valuable capabilities that might assist them hold their brains sharp. They can help adults, too, with the useful resource of manner of retaining their capabilities honed and their brains functioning optimally.

Alphabet Categories

You can play this interest with only one little one or with a collection of youngsters. It moreover may be excellent a laugh to play with humans of mixed ages. One participant alternatives a awesome class, in conjunction with automobiles, animals, or tv indicates. The first player names 5 of some thing that falls in that magnificence and begins offevolved with the letter "A." The second participant then does the equal, the usage of the letter "B." Players maintain to take turns down the letters of the alphabet till a person gets caught. They then can alternate to a terrific elegance and begin over.

Card Memory

The object of this undertaking is to test one's reminiscence and attention abilities. It can be done solo or with more than one gamers. One participant shuffles a deck of playing playing cards after which lays each card out face down in a rectangular grid sample. Each participant takes turns flipping over playing cards of his or her choice. If the gambling gambling playing cards are an same pair, the player gets to preserve the gambling playing cards and turns over some different pair of gambling gambling cards. If the gambling cards are not an identical pair, all contributors attempt to bear in mind what they're; then the cards are grew to emerge as face down another time, and the following player takes a flip. The game continues until all pairs were matched. The player with the maximum playing cards at the quit is the winner.

Color Cars

This is a a laugh sport to play when you're using inside the car alongside facet your children. It moreover enables extra younger youngsters take a look at colors and counting. Before commencing on an errand, every player selections a color after which guesses how many cars of that colour she or he could be capable of see at the way for your excursion spot. Each player is responsible for preserving depend of what number of automobiles he or she sees within the color he or she picked. The player with the exceptional rely while you attain your excursion spot is the winner. On an extended street experience you can divide this sport into smaller intervals of 10 mins, or prevent on the identical time as a person has visible, say, 20 cars in a single colour.

General Trivia

Trivial Pursuit is a well-known and famous board recreation that exams the overall records of its game enthusiasts. However, in case you don't own the sport, you could with out problems do a short on line are searching out that yields lots of trivialities questions with solutions and use them to make your very non-public version for pals and own family to play.

One sixth-grade instructor prepares 10 questions after which asks his college students to jot down down their answers to every query on a sheet of coated paper. Afterwards, they evaluation the answers together to appearance who were given the maximum proper. This may be a fun party recreation as properly, with prize trinkets for the humans who've the most correct answers.

Hangman

This simple game, requiring brilliant some thing to put in writing with and some trouble to install writing on, is a incredible way to test phrases. One player secretly selects a word and then writes out one location for every letter within the word. The particular participant has to guess the phrase through randomly guessing man or woman letters. If the letter belongs inside the word, then the number one player writes the letter in the blanks wherein it takes location. If the player options a letter that isn't in the word, the opposite participant will draw a frame detail on the hangman. If the hangman is entire and the word isn't in reality spelled out, the game is misplaced.

This activity can cross on for quite a while, relying on how unique you want to get collectively together with your hangman. Some youngsters want to keep it simple with the resource of completing with a head, frame, palms and two legs but you could keep the sport going as long as you want via using adding hands, hair, shoes, clothes and so forth.

I Spy and I Rhyme

This exercise is much like the traditional "I Spy" game however has a little bit of an academic twist. The cause of I undercover agent and I rhyme is to discover an object inside the room that rhymes with the item the preceding participant has decided on. For example, if player one says "I spy a book," then participant may additionally additionally say "I secret agent a hook" or "I undercover agent a corner."

License Plate Addition

This fun math sport may be played at the same time as the use of or even as taking a stroll round a network that has many parked vehicles. One participant appears at a license plate range after which calls out the numbers which is probably on it. Most license plates include to 3 numbers. The first player who can efficiently add those numbers wins. With higher-professional gamers, you may add numbers of or 3 digits collectively or multiply single digits.

Step School

If you have a staircase in your home you could attempt gambling "step university." The backside step is considered the bottom grade possible and the pleasant step is the very quality schooling viable. Everyone begins offevolved offevolved at the bottom step and an older toddler or adult serves as the teacher. The trainer asks all people a query and the difficulty is primarily based totally on what step that player is on. If a player is on the bottom step, the instructor might probably ask an clean question together with "What is 1+1?" and then asks increasingly more harder questions. Whoever have to make it to the top step first is the winner.

The States Game

This is some exceptional exquisite task that can be accomplished nearly anywhere and it is also very instructional. One member of the family begins out through the usage of the usage of choosing a rustic. The subsequent player has to call a state that begins offevolved offevolved with the closing letter of that u . S . A .. If Mom alternatives Georgia, Dad may additionally pick Alaska, Junior would likely pick Arkansas, and so forth. You can also do that for towns, hues, and so on.

Traveling Word Alliterations

This challenge is high-quality for exciting the minds of children and training them approximately phrase alliteration. The intention of the sport is for each player to assume that she or he may be taking a experience somewhere and could want to carry some thing that begins offevolved as the identical letter as their tour spot. For instance, a participant can say "I'm visiting to Florida and bringing my Fish," or "I'm travelling to Cousin Charlie's and bringing my Crayons." To make the game greater fun, encourage creativity and silliness.

War

This a laugh card activity requires a deck of playing cards and at least gamers. It can assist children research the order of numbers and their relative values.

One player shuffles the deck of playing playing cards and offers the entire deck out face down. Once every player has a pile of cards, all gamers turn over their top card. Whoever has the card with the very excellent variety wins all of the cards. For example, if one player flips over a ten and the second one participant flips over a 3, the number one player, having the very best issue fee, may additionally collect both playing cards. If every gamers pull an same massive variety, each participant draws 3 playing gambling playing cards, preserving them facedown and then flips over the fourth card. Whoever has the satisfactory-valued card receives all the playing playing playing cards. The first player to collect all the playing playing cards is the winner.

Chapter 6: World History To Remember

"Those who do now not study data are doomed to duplicate it."

1. What u.S. Of the usa of america's monarchy became overthrown all through the 14 july Revolution? Iraq

2. What House of Plantagenet department did House of York fight in competition to in the War of the Roses? House of Lancaster

3. What excessive-ranking Nazi decent became assassinated as the result of Operation Anthropoid? Reinhard Heydrich

4. To what island became Napoleon Bonaparte exiled to in 1814? Elba

5. What is the decision of the revolution that overthrew China's Qing dynasty in 1911? Xinhai Revolution

6. Who changed into the partner to King Ferdinand II of Aragon? Isabella I of Castile

7. In what u . S . Did the 1937 Parsley Massacre rise up? Dominican Republic

8. On what river changed into the historic metropolis of Babylon built? Euphrates

nine. Who initiated the Salt March in 1930? Mahatma Gandhi

10. Who changed into Chairman of the Provisional Government of the Irish Free State? Michael Collins

11. Who succeeded Benito Mussolini as Prime Minister of Italy? Pietro Badoglio

12. Who modified into First Lady of Argentina on the time of her loss of life in 1952? Eva Perón

13. How many prison recommendations consist of the Code of Hammurabi? 282 legal guidelines

14. What European king founded the Congo Free State? Leopold II of Belgium

15. In what u.S. Of the us did the Boxer Rebellion take vicinity? China

16. What faction did Alexander Bogdanov co-located? Bolsheviks

17. Who emerge as the Archbishop of Canterbury that have become murdered in 1170? Thomas Becket

18. Mustafa Atatürk changed into the number one President of what america of the united states? Turkey

19. D. F. Malan started instituting apartheid as Prime Minister of what u.S.A.? South Africa

20. Jochi end up the eldest son of what ruler? Genghis Khan

21. Who have become the number one Tsar (Czar) of Russia? Ivan IV

22. During what conflict grow to be the Battle of the Somme fought? World War I

23. Who led the coup in competition to Milton Obote and characteristic emerge as President of Uganda in 1971? Idi Amin

24. Airlifts of food and assets started out arriving to what European town on June 26, 1948? Berlin

25. In what Central American u . S . Did the Contras conflict the Sandinistas? Nicaragua

26. In what 12 months did Mount Vesuvius erupt and bury Pompeii? Seventy nine A.D.

27. What ruler suffered his satisfactory defeat at Battle of the Catalaunian Plains? Attila the Hun

28. It is concept that Machu Picchu became built as an property for what Inca emperor? Pachacuti

29. In what united states of america did Mulai Ahmed er Raisuli and his bandits kidnap Ion Perdicaris and his step-son in 1904? Morocco

30. Who led the twenty 6th of July Movement? Fidel Castro

31. Traudl Junge turned into a private secretary to what dictator? Adolf Hitler

32. Adrian IV is the exceptional Pope to had been born in what European u . S . A .? England

33. King Mongkut changed into king of what modern-day u . S . At a few thing of the 1800's? Thailand (previously Siam)

34. In what metropolis did the Tiananmen Square protests of 1989 occur? Beijing, China

35. Who have become the remaining leader of the Soviet Union? Mikhail Gorbachev

36. The throne of the Holy Roman Empire changed into continually captivated with the resource of what royal house from 1438 to 1740? House of Habsburg (Also known as House of Austria)

37. Who grow to be the Prime Minister that Yigal Amir assassinated on November four, 1995? Yitzhak Rabin

38. Darien colony have emerge as based by the use of the Kingdom of Scotland at the land of what modern-day u.S.A. Of america in 1698? Panama

39. What became the capital of the Old Kingdom of Egypt? Memphis

forty. Who lead the number one excursion to achieve the South Pole? Roald Amundsen

41. Who succeeded Winston Churchill as Prime Minister of the United Kingdom in 1945? Clement 1st Earl 1st earl attlee

40 . Who turn out to be the very last Aztec Emperor? Cuauhtémoc

40 three. What is the call of the dynasty that became overthrown within the path of the 1979 Iranian Revolution? Pahlavi dynasty

44. Kingdom of Gorkha is the preceding call for what current day Asian u . S . A .? Nepal

45. What emerge as the decision of the area's first synthetic nuclear reactor? Chicago Pile-1

46. What New7Wonders of the World place modified into as soon as the capital of the Nabataeans? Petra

forty seven. Who have end up elected the primary woman Chancellor of Germany? Angela Merkel

forty eight. What is the call of the castle that the shogun of the Tokugawa shogunate of Japan as soon as referred to as domestic? Edo Castle

forty 9. Who emerge as the primary black Archbishop of Cape Town, South Africa? Desmond Tutu

50. What is the call of the female who purposely stepped in the front of King George

V's horse on the Epsom Derby on June four, 1913? Emily Davison

51. Who did the Jordanian Armed Forces struggle in the course of Black September in 1970? Palestine Liberation Organization

fifty. Who end up chief of the Partisans all through World War II? Tito

fifty 3. What u . S . A . Did Henry Morgan grow to be Lieutenant Governor of? Jamaica

fifty 4. In what town changed into Archduke Franz Ferdinand of Austria assassinated? Sarajevo

fifty five. Anschluss have become the name of the annexation of what u.S.A. Of america in Nazi Germany? Austria

56. What Spanish conquistador captured and killed Incan emperor Atahualpa? Francisco Pizarro

57. The United States' invasion of what small state become codenamed Operation Urgent Fury? Grenada

fifty 8. Who advanced the smallpox vaccine? Edward Jenner

fifty nine. Who commanded all Allied ground forces at some point of Operation Overlord (Battle of Normandy)? Field Marshal Bernard Sir Bernard Law

60. Nicolae Ceaușescu have come to be the final Communist chief of what u . S .? Romania

sixty one. Who based totally Quebec City? Samuel de Champlain

62. How many inmates had been imprisoned on the Bastille while it changed into stormed on July 14, 1789? Seven

sixty 3. Who became the primary European to discover the ocean route to India? Vasco da Gama

sixty four. Treaty of Nerchinsk set the border among Russia and what different usa of america? China sixty five. Van Diemen's Land grow to be the unique call Europeans used for what island? Tasmania

66. In what year changed into the Great Fire of London? 1666

sixty seven. Who changed into looking for the Seven Cities of Gold at the same time as he positioned the Grand Canyon? Coronado

sixty eight. What 20th century struggle is also referred to as the Fatherland Liberation War? Korean War

sixty nine. In what African u . S . A . Did the War in Darfur take place? Sudan

70. In what u.S. Of the united states of the united states turned into Leon Trotsky assassinated in 1940? Mexico

71. What have become the number one u.S. To enact ladies's suffrage? New Zealand

seventy two. Who ran the number one marathon among Marathon and Athens, Greece? Pheidippides seventy three. In what united states of america changed into the Maginot Line constructed? France

seventy four. What well-known Russian did Felix Yusupov help assassinate in 1916? Grigori Rasputin

seventy five. In what u.S.A. Did the Carlist Wars get up at a few degree inside the nineteenth century? Spain

76. Who primarily based the Mughal Empire? Babur

77. In what u.S.A. Of america did the Blackshirts shape in 1923? Italy

seventy eight. Who have become the mom of Queen Elizabeth I? Anne Boleyn

79. What become the call of the Hungarian Countess who changed into attempted and convicted of a couple of murders in 1611? Elizabeth Bathory

80. Who captained the second a fulfillment circumnavigation of the arena? Francis Drake

eighty one. What present day international locations have been fashioned because of the Velvet Divorce? Czech Republic and Slovakia

eighty two. What explorer claimed Canada for France? Jacques Cartier

eighty three. In what English city did the Peterloo Massacre stand up on August 16, 1819? Manchester

84. What treaty granted Mexico its independence from Spain? Treaty of Córdoba

eighty five. In what u . S . A . Did the Rose Revolution stand up in 2003? Georgia

86. Who received the Orteig Prize in 1927? Charles Lindbergh

87. In what present-day usa did the Maji Maji Rebellion arise inside the early twentieth century? Tanzania

88. What monarch did Alexandra Feodorovna marry in 1894? Nicholas II of Russia

89. In what usa of the united states did the 1916 Easter Rising (Easter Rebellion) occur? Ireland

90. Who is the Scandinavian that based the primary Norse agreement on Greenland? Erik the Red

ninety one. Who modified into the fifth Great Khan of the Mongol Empire? Kublai Khan

90. On what island u.S. Changed into Ferdinand Magellan killed at the same time as trying to circumnavigate the area? Philippines

ninety three. What is the choice of the British ocean liner did German U-boat SM U-20 sink on May 7, 1915? RMS Lusitania

90 4. Francisco de Orellana changed into the primary character to efficaciously navigate what river? Amazon

90 five. In what Pakistan city have become Osama bin Laden observed and killed? Abbottabad

ninety six. What is the decision of the big explosion that passed off in Siberia on June 30, 1908? Tunguska Event

90 seven. What metropolis-nation did Athens fight in opposition to within the Peloponnesian War? Sparta

ninety eight. What is the name of the British nurse who changed into tried for treason and performed in 1915? Edith Cavell

ninety nine. Who changed into the primary Pope to go to North America? Pope Paul VI (1965)

one hundred. What became the primary Naval struggle wherein plane providers engaged each other? Battle of the Coral Sea (WWII)

one zero one. By what name is the Anti-Fascist Protection Rampart better called? Berlin Wall

102. What European monarch died on September 14, 1982? Grace Kelly

103. What did the Nimrod Expedition find out? Antarctica

104. What piece of art work did Vincenzo Peruggia thieve on August 21, 1911? Mona Lisa

one zero five. In what u . S . A . Changed into Dolly the cloned sheep born in 1996? Scotland

106. In what town and u . S . A . Changed into the capital of the League of Nations located? Geneva, Switzerland

107. During what naval conflict have become Admiral Horatio Nelson killed? Battle of Trafalgar

108. What is the decision of the Roman preferred who devoted suicide after losing the Battle of Actium? Mark Antony

109. What america of america did Egypt declare its independence from in 1922? United Kingdom

one hundred ten. In what current-day u.S.A. Did the Boer Wars get up? South Africa

111. The death of what character result in the 2011 English riots? Mark Duggan

112. How many famous Roman Catholic Popes served for the duration of the twentieth Century? Nine

113. Where did Curiosity land in 2012? Mars

114. Who is known as "The Nazi who said sorry"? Albert Speer

one hundred fifteen. What is the decision of the previous Australian Prime Minister that went lacking and end up later presumed dead in 1967? Harold Holt

116. What German physicist became the originator of quantum idea? Max Planck

117. Who emerge as Chief of State of Vichy France (Free State) at some stage in World War II? Philippe Pétain

118. Who modified into the excellent survivor of the auto crash that killed Diana, Princess of Wales in 1997? Trevor Rees-Jones

119. Mary Mallon is idea to have unfold what infectious illness? Typhoid (Typhoid Mary)

100 and twenty. What form of object is a Fokker Eindecker? Fighter aircraft

121. Who is appeared because the inventor of current paper? Cai Lun

122. Benjamin Guinness come to be the primary Lord Mayor of what capital city? Dublin, Ireland

123. What World War II operation turn out to be nicknamed Dynamo? Evacuation of Dunkirk

124. Who overthrew Libya's King Idris I in 1969? Muammar Gaddafi

100 twenty five. Cedric Popkin is the person believed to have shot and killed what

World War I fighter? Manfred von Richthofen (Red Baron)

Science & Nature Up Close

Science is Everywhere

1. What is the rarest human blood kind? AB poor

2. What constellation's call in Latin manner "lizard"? Lacerta

three. Where might you visit locate the area's most atmospheric strain? Siberia

4. After hydrogen, what's the second maximum enough gas that forms the Sun? Helium

five. What type of galaxy is the Milky Way? Spiral galaxy

6. A halophile organism needs a excessive interest of what compound to live to inform the story? Salt

7. How many tiers are in a heptagon? 900 tiers

8. What are the primary 7 digits of Pi? Three.141592

nine. What sort of pancreatic cells produce insulin? Beta cells

10. What is the maximum huge blood protein located in people? Albumin

eleven. One unit of horsepower is same to approximately what number of watts? 746 watts

12. What is the choice for the look at of substances at very low temperatures? Cryogenics

13. What planet in our sun gadget has the most recognized satellites? Saturn

14. In what part of the mobile is DNA discovered? Nucleus

15. What is the best clearly taking area fissile isotope? Uranium-235

16. What sickness does someone have if they'll be recognized with urolithiasis? Kidney stones

17. Does lightning excursion at the price of light? No

18. What is the most important three-digit high amount? 997

19. What shape of tool is a Pascaline? Calculator

20. What does pH stand for? Potential of hydrogen

21. Which planet has the moon Callisto? Jupiter

22. Aqua fortis is more usually called what acid? Nitric acid

23. What form of cloud produces rain? Nimbus

24. How many vertices does a dice have? Eight

25. Pi instances the radius squared solves for the vicinity of what? Area of a circle

26. Which is a higher conductor of power: gold, copper or silver? Silver

27. Dromedary is a sort of what mammal? Camel

28. What changed into the primary infectious contamination to be eliminated? Smallpox

29. Which part of the thoughts produces oxytocin? Hypothalamus

30. What is the lightest steel? Lithium

31. What kind of compound can act as every an acid and a base? Amphoteric

32. Landspout and waterspout are styles of what climate event? Tornados

33. How many protons are in a single oxygen atom? Eight

34. What does the acronym PVC stand for in PVC pipe? Polyvinyl chloride

35. What changed into the primary spacecraft to land on the Moon? Luna 2 (Lunik 2)

36. What is nomophobia the priority of? Fear of being out of mobile phone contact

37. How many exclusive forms of nutrients are within the human body? Thirteen

38. A barn is a unit of what? Area

39. What "royal water" can dissolve each gold and platinum? Aqua regia

forty. What is the world's largest flower? Rafflesia arnoldii

forty one. Which element has the chemical symbol P? Phosphorous

42. How many organizations are on the Periodic Table of Elements? 18

forty 3. In which layer of Earth's surroundings is the ozone layer? Stratosphere

44. What is the nearest-identified famous man or woman to the Sun? Proxima Centauri

forty five. What is diaphoresis? Perspiration

forty six. How many favored cubic ft is a standard twine of timber? 128 cubic toes

forty seven. Poisoning from what metallic motives Itai-itai sickness? Cadmium

forty eight. What is the pH of healthful spinal fluid? 7.Four

49. How many flavors of quarks are there? 6

50. Hippology is the take a look at of what mammal? Horses

fifty one. Which planet's one year is shorter than its day? Venus

fifty . What is the International Systems of Units (SI) unit used frequently in chemistry for quantity of substance? Mole

fifty three. What metallic detail is copper blended with to create bronze? Tin fifty four. What form of clouds are the very best in Earth's environment? Noctilucent clouds

fifty 5. What kind of substance motives maximum cancers? Carcinogen

fifty six. What poison is used on South American poison darts? Batrachotoxin

57. What theorem states "A squared plus B squared equals C squared"? Pythagorean theorem

fifty eight. What element become formally called hydrargyrum? Mercury

fifty nine. Is crude oil an acid or is it a base? It's neither

60. What is the simplest letter not used at the Periodic Table? J

61. What mammal is known to stay the longest? Bowhead whale

sixty . Where is the area's largest series of corium placed? Chernobyl

sixty 3. How many grams are in a decagram? 10

sixty four. What is considered the simple unit of life? The cellular

65. What protein is used to make gelatin? Collagen

Chapter 7: What Type Of Whale Can Grow A Prolonged Tusk?

a hundred and one. What is the strong shape of carbon dioxide called? Dry ice

102. What type of virus is the reason for the common bloodless? Rhinovirus

103. How many teaspoons of water might not it take to fill a gallon? 768 teaspoons

104. Equinoxes arise every 365 days inside the path of what months? March and September

one 0 five. How many classes of hurricanes are there? Five

106. To the nearest ten-million, what number of miles is Earth's whole rotation at some stage in the Sun? 584 million miles

107. What is the call for a solution wherein the attention of solutes is extra outside the cell than indoors it? Hypertonic

108. Sodium and potassium belong to which agency on the Periodic Table? Alkali metals

109. How many protons does an alpha particle comprise? 2

110. How many spiral fingers does the Milky Way have? Four

Geography

Here, There and Everywhere

1. After London, what's the UK's second most populated metropolis? Birmingham, England

2. What frame of water separates Asian Turkey from European Turkey? Bosphorus

three. After China, India and the united states of America, what's the arena's fourth maximum populated usa? Indonesia

four. What frame of water separates Russia and Alaska? Bering Strait

five. What is the Arctic Ocean's biggest island? Victoria Island (Canada)

6. In what metropolis may additionally you walk all through the Pont Neuf? Paris, France

7. Alphabetically, what African capital metropolis is very last? Yaoundé, Cameroon

8. Inside what present day-day u . S . Is the historical town of Carthage? Tunisia

nine. In what u.S. May additionally need to you discover the intact remnants of Hadrian's Wall? England

10. What volcanic explosion is taken into consideration the loudest sound ever heard in modern facts? Krakatoa

11. What is the sector's tallest brick building? Chrysler Building

12. What sea does Croatia border? Adriatic Sea

13. Flodden Wall turn out to be constructed to help defend what European city? Edinburgh, Scotland

14. What Greek temple became constructed in self-discipline to the goddess Athena? Parthenon

15. How many rooms are inside the White House? 132

16. What lake was customary because of the fall apart of Mount Mazama? Crater Lake

17. Republic of Macedonia emerge as created after the breakup of what usa? Yugoslavia

18. What is Europe's longest mountain chain? Scandinavian Mountains

19. How most of the arena's nations start with Q? One (Qatar)

20. What is the official language of Cambodia? Khmer

21. What is the most visited museum within the worldwide? Forbidden City

22. What is the largest hydraulically stuffed dam in the United States? Fort Peck Dam

23. After Mount Kilimanjaro, what is the second tallest mountain in Africa? Mount Kenya

24. What is the tallest freestanding shape in North America? CN Tower (Toronto)

25. The Matterhorn sits the various border of Switzerland and what one among a kind america? Italy

26. In which ocean is Chukchi Sea located? Arctic Ocean

27. Where should you find out the area's largest energetic geyser, Steamboat Geyser? Yellowstone National Park

28. What is the biggest city positioned sincerely at the European continent? Moscow

29. In what u . S . A . Could you discover the ruins of Tikal? Guatemala

30. What gulf separates Yemen and Somalia? Gulf of Aden

31. What is the area's tallest form? Burj Khalifa (Dubai)

32. What is the name of the Mexican volcano that commenced erupting in 1943? Paricutin

33. What river is 4,258 miles lengthy? Nile

34. What is the area's steepest residential road? 1st Earl 1st Earl Baldwin of Bewdley of Bewdley Street (Dunedin, New Zealand)

35. What is the last volcano at the European mainland to erupt? Mount Vesuvius

36. What is the largest island in the Caribbean? Cuba

37. To the closest a hundred,000 miles, how many square miles is the land area of america? Three,537,438 square miles

38. Which U.S. Country best has 3 counties? Delaware

39. In what European metropolis is The Little Mermaid statue? Copenhagen, Denmark

40. What is the most important island within the Southern Hemisphere? New Guinea

forty one. The Bechuanaland Protectorate have become what u.S.A. In 1966? Botswana

forty. What World Heritage Site may be located in Wiltshire, England? Stonehenge

forty 3. What is the center of the Northern Hemisphere? North Pole

44. What fountain sits in the front of the Palazzo Poli? Trevi Fountain

45. What u . S . Is home to Ngorongoro Crater? Tanzania

forty six. What fortress emerge as used as the principle film area for the television collection Downtown Abbey? Highclere Castle

forty seven. How many recollections tall is the Empire State Building? 102

forty eight.   In what European united states of the usa might you discover Po Valley? Italy

forty nine.   What is the sector's largest museum? Louvre

50.   What historical landmark may need to you be journeying in case you went to La Cuesta Encantada (The Enchanted Hill)? Hearst Castle

51.   What is the very great peak inside the Cascade Range? Mount Rainier

fifty . In which U.S. State may also go to visit Carlsbad Caverns? New Mexico

fifty three.   What is the most crucial of the three waterfalls that shape Niagara Falls? Horseshoe Falls

fifty four.   What u . S . Is home to the westernmost problem in Europe? Portugal

fifty 5. What is the real language of Greenland? Greenlandic fifty six. What is the maximum densely populated South American u.S.A. Of the usa? Belize

fifty seven. What continent is domestic to Alice Springs? Australia

fifty eight. In what gulf is Bight of Bonny placed? Gulf of Guinea

59. What European the usa of the us is domestic to peace strains (walls) which have been built to cut up Catholics and Protestants? Northern Ireland

60. After the Dead Sea, what is the second one-lowest lake in the global? Sea of Galilee

sixty one. Along with Spain and France, what is the most effective super u . S . To have every Mediterranean Sea and Atlantic Ocean coastlines? Morocco

sixty . What is the best landlocked united states in Southeast Asia? Laos

63. What is the most crucial territory of the USA? Puerto Rico

sixty 4. What European city is home to the area's largest ancient fortress? Prague (Prague Castle)

sixty five. What is the most vital town in the French Riviera? Nice, France

66. In what u . S . Are the Nazca Lines located? Peru

sixty seven. Puget Sound belongs to what sea? Salish Sea

68. In what U.S. National Park should you locate Half Dome? Yosemite National Park

sixty nine. Kariba Dam shares its place with Zambia and what special African america? Zimbabwe

70. What is the smallest African u . S . A .? Seychelles

71. How many miles prolonged is the Panama Canal? Forty 8 miles

seventy two. In what united states of america may need to you find out remnants of the ancient city of Troy? Turkey

seventy 3. How many landlocked international locations are there in North America? Zero

seventy four. Natives of what African u . S . A . Are known as "Cha-bos"? Chad

seventy five. Hook Lighthouse is the oldest running lighthouse in what u.S.? Ireland

seventy six. What is the southernmost variety wherein the Sun may be right now overhead? Tropic of Capricorn

77. What is the area's longest river to go along with the float totally within one u . S .? Yangtze (China)

seventy eight. What non-European u . S . A . Has the most Dutch talking residents? Suriname

seventy nine. What is Europe's most moderately populated u.S.A.? Iceland

80. What become the capital of West Germany? Bonn

80 one. What U.S. Country capital has the pleasant elevation? Santa Fe, New Mexico

eighty two. What is the area's driest non-polar barren location? Atacama Desert in South America

80 3. What parallel separates Canada and Montana? Forty ninth parallel

84. Gotland is an island that belongs to what u . S . A .? Sweden

eighty 5. What is the most populated city in Florida? Jacksonville

86. What is the inner maximum issue in Earth's seabed? Challenger Deep

87. What is the area's largest enclave landlocked u.S. Of the us? Lesotho

88. What is the arena's maximum mountain out of doors of Asia? Aconcagua (Argentina)

89.   Aside from the oceans, what's the area's largest biome? Taiga

90.   What is Siberia's highest peak? Mount Belukha

ninety one.   The Euxine Sea is better regarded through manner of what call? Black Sea

90 .   Which is the least populated of New York City's boroughs? Staten Island

ninety three. What continent has the very super not unusual elevation? Antarctica

ninety four.   Volcán Wolf is the very excellent issue on what island enterprise? Galapagos Islands

ninety five.   What is the area's northernmost metropolitan city? Helsinki, Finland

96.   How many U.S. States have been completely or in part created thru the Louisiana Purchase? 15

90 seven.   What is the poorest u . S . A . In Europe? Moldova

98.   What is the very quality summit of the Rocky Mountains? Mount Elbert

99.   What capital metropolis modified its name from Edo in 1868? Tokyo

a hundred.   What is the northernmost country within the continental United States? Minnesota

101.   What is the handiest survivor of the Seven Wonders of the Ancient World? Great Pyramid of Giza

102.   What is the pleasant capital city in Europe? Andorra l. A. Vella

103.   What U.S. Rock formation became inside the starting called Temple of Aeolus? Angels Landing

104.   What is the elevation of Antarctica on the South Pole? Nine,three hundred toes

one zero five. What modified into the primary 8,000 meter (26,two hundred ft) mountain to be summited? Annapurna I (Nepal)

106. What is the most densely populated island city in North America? Manhattan, NY

107. What worldwide landmark is placed within the city of Agra? Taj Mahal

108. Into what frame of water does the Euphrates river drain? Persian Gulf

109. What is the decision of the non-public limestone cave within the United States? Tears of the Turtle Cave

a hundred and ten. In what u.S.A. Is the rock castle Sigiriya positioned? Sri Lanka

111. What continent has the maximum landlocked international locations? Europe

112. What South American region encompasses the location's largest tropical wetland place? Pantanal

113. To the closest 1 million, how many square miles is the world's stylish land vicinity? Fifty seven,393,000 square miles

114. Which U.S. Country is most in propose elevation? Colorado

115. What are the handiest double landlocked international places inside the worldwide? Bhutan and Uzbekistan

116. The biggest a part of the Kalahari Desert is situated indoors what African country? Botswana

117. What is the non-public lake in North America? Great Slave Lake

118. After Cairo, what's the second one largest city within the Arab global? Baghdad

119. In what town and nation is the Thaddeus Kosciuszko National Memorial? Philadelphia, Pennsylvania

Chapter 8: What River Feeds Grand Canyon Of The Yellowstone?

121. Surtsey is the southernmost component of what america? Iceland

122. How many mountain peaks have an elevation better than 28,000 ft? Three

123. Besides the Blue Nile, what is the opportunity vital tributary of the Nile? White Nile

124. What shape of glacier is a Piedmont glacier? Valley

a hundred twenty 5. Bab-el-Mandeb (Gate of Tears) connects Gulf of Aden to what sea? Red Sea

126. What usa's most mountain is Snowdon? Wales

127. Where is the biggest public bison herd within the U.S.? Yellowstone National Park

128. What is the maximum vital landlocked u . S . In South America? Paraguay

129. Argentina, Chile and what different u . S . Comprise the Southern Cone of South America? Uruguay

a hundred thirty. In what Canadian territory is the Klondike vicinity? Yukon

131. Aotearoa is the Māori call for what u . S .? New Zealand

132. Split is the second-largest town in what European u.S.A. Of the us? Croatia

133. What is the southernmost u . S . A . In New England? Connecticut

134. How many countries border each China and Russia? 14

130 5. Which Seven Wonders of the Ancient World grow to be destroyed thru the 1303 Crete earthquake? Lighthouse of Alexandria

136. In what U.S. Kingdom is Bad Rock Canyon located? Montana

137. What is the most populous metropolitan region within the global? Tokyo, Japan

138. What is both the very awesome pinnacle in Mexico and the brilliant volcano in North America? Pico de Orizaba

139. What is the most densely populated U.S. Country? New Jersey

a hundred and forty. In what Asian capital metropolis is Chiang Kai-shek Memorial Hall? Taipei, Taiwan

141. Which is the southernmost Scandinavian u . S .? Denmark

142. What is the reliable language of Ivory Coast? French

143. Which united states is extra in basic place: Italy, Spain or Germany? Spain

one hundred forty four. What the usa is positioned closest to Antarctica? Argentina

145. In which ocean is the Ring of Fire? Pacific Ocean

146. What U.S. Kingdom is nicknamed The Diamond State? Delaware

147. Which Great Lake is Chicago located on? Lake Michigan

148. What U.S. State is home to Naval Base Kitsap? Washington

149. What is Maine's maximum populated town? Portland

150. Rabat is the capital of what u . S . A .? Morocco

Sports Junkie

Take it to the house

1. Who is the oldest character to ever rating an NFL touchdown (non-throwing)? Doug Flutie

2. How many universal groups had been in Major League Baseball while the Chicago Cubs received the 1908 World Series? Sixteen

three. What does GSH stand for on the uniforms of the Chicago Bears? George Stanley Halas

four. What North American football club did Pelé play inside the course of the 1970's? New York Cosmos

five. What have become the genuine call of the National Football League in advance than changing its name in 1922? American Professional Football Association

6. Actor Mark Harmon's dad Tom Harmon acquired the Heisman Trophy at the same time as playing for what college? University of Michigan

7. What is the quality finish for the united states guys's institution within the FIFA World Cup? Third Place (1930 World Cup)

eight. Who end up the fine fighter to knockout Floyd Patterson in consecutive bouts? Sonny Liston

9. What is the ultimate occasion of a decathlon? 1,500-meter run

10. Who changed into the primary participant to sign with Major League Soccer? Tab Ramos

11. Who changed into the number one San Francisco 49er to win NFL MVP? John Brodie

12. From what u . S . A . Does MMA fighter Conor McGregor hail from? Ireland

13. What u.S.A. Turned into first of all decided on to host the 1986 FIFA World Cup? Colombia

14. Who modified into the primary winner of the Hickok Belt? Phil Rizzuto

15. Who is the great-ever leader in passing yards for the USFL? Bobby Hebert

sixteen. What did Hans-Gunnar Liljenwall emerge as the first athlete to fail? Olympic drug take a look at

17. What heavyweight boxing champion come to be nicknamed Boston Strong Boy? John L. Sullivan

18. Who emerge as the primary father-son duo to hit once more-to-back domestic runs in an MLB exercise? Ken Griffey Jr. And Sr.

19. Who changed into the number one player to file a triple-double within the NBA All-Star Game? Michael Jordan

20. Who became the final NFL player to play a recreation without a helmet? Dick Plasman

21. Who is the most effective MLB player to win a Gold Glove Award as each an outfielder and as an infielder? Darin Erstad

22. Who is the brilliant male wild card entrant to win a Wimbledon pick out out? Goran Ivanišević

23. What is the decision of the athletic mascot for the University of Texas at Austin? Bevo

24. Who was supervisor of the Birmingham Barons the season Michael Jordan finished for them? Terry Francona

25. What NFL crew did Vince Lombardi take over as head instruct and well-known manager for after leaving the Green Bay Packers? Washington Redskins

26. Who have become the first lady Naismith College Player of the Year? Anne Donovan

27. How many minutes is a golfer allowed to search for a out of region ball? Five-mins

28. What became the genuine call of Wrigley Field? Weeghman Park

29. What does BMX (biking) stand for? Bicycle moto x

30. As of 2017, who's the closing undisputed heaving boxing champion? Lennox Lewis

31. What racing event continually takes location the primary Saturday in May? Kentucky Derby

32. What nations done in the first cricket take a look at healthful? Australia and England

33. What became the number one African us of a to qualify for the FIFA World Cup? Egypt

34. Who is the simplest unique organizations player to win NFL MVP? Mark Moseley (1982)

35. Who modified into the first tennis participant to win the Golden Slam with the useful resource of triumphing all four Grand Slam singles titles and the Olympic gold medal within the same calendar yr? Steffi Graf

36. What emerge as the number one NFL institution to win 5 Super Bowls? San Francisco 49ers

37. Who become the primary American to manipulate an English Premier League club? Bob Bradley

38. Who batted inside the decrease back of Joe DiMaggio even as DiMaggio made his Major League Baseball debut? Lou Gehrig

39. Who is the first-class head train to win three Super Bowls with three special starting quarterbacks? Joe Gibbs

40. Who changed into the primary New York Knick to guide the NBA in scoring? Bernard King (1984-eighty 5)

41. What jersey massive variety did Michael Jordan located on whilst he un-retired and rejoined the Chicago Bulls? Forty 5

40. Where did Dwayne "The Rock" Johnson play college football? University of Miami

40 3. Who have emerge as the primary MLB pitcher to hit home runs in the equal sport he threw a no-hitter? Rick Wise

44. What Brazilian soccer club tragically misplaced 19 players in a 2016 aircraft crash? Chapecoense

forty five. The global hall of reputation for what recreation is in Canastota, New York? Boxing

46. Who become the primary player to attain 4 goals in his NHL debut? Auston Matthews

forty seven. How many corporations from the American Basketball Association joined the NBA following the ABA-NBA merger in 1976? Four

48. Who is the English Premier League's all-time important purpose scorer? Alan Shearer

40 9. Who have end up the number one tennis player to earn one-million dollars in a unmarried season? Bjorn Bjorg

50. Who modified into the number one MLB participant to record four,000 regular season hits? Ty Cobb

51. When modified into the final time England acquired the FIFA World Cup? 1966

fifty . Who is the last NBA player to not unusual 30 factors and 15 rebounds in line with game for an entire season? Bob McAdoo

fifty three. What professional football membership signed Tim Howard in 2003? Manchester United

54. Who changed into the primary head educate for the Jacksonville Jaguars? Tom Coughlin

fifty five. Who holds the Major League Baseball record for max video games completed without appearing in a World Series? Rafael Palmeiro fifty six.   What NBA team modified into Dick Vitale head train of in the past due 1970's? Detroit Pistons

fifty seven.   Who modified into the number one non-quarterback or non-jogging again to win NFL MVP? Alan Page

58.   Who become the first man or woman inducted into the Basketball Hall of Fame as each a participant and teach? John Wooden

59.   Who end up the number one U.S. President to seem on the cover of Sports Illustrated? John F. Kennedy

60.   Who end up the primary soccer participant to win four European Golden Shoe awards? Cristiano Ronaldo

sixty one.   What organization acquired the very last USFL Championship? Baltimore Stars

sixty two.   What cutting-edge NBA crew have become as quickly because the Buffalo Braves? Los Angeles Clippers sixty 3.   Who is the youngest winner of a Grand Slam singles become privy to? Michael Chang

sixty four.   Ryan Giggs and Paul Scholes executed their complete careers for which English Premier League membership? Manchester United

sixty five.   How many furlongs is the Kentucky Derby? 10

66.   What Pro Football Hall of Famer changed into born in Port Angeles, WA? John Elway

67.   Who have become the primary North American to win a expert sports activities activities championship as a player, teach, and authorities? Pat Riley

68.   What NFL quarterback became suspended due to "Deflate Gate"? Tom Brady

sixty nine.   Who holds the NBA record for max career assists? John Stockton

70. What was the name of the unique domestic field for the Pittsburgh Steelers? Forbes Field

seventy one. Who gave Evander Holyfield his first expert loss? Riddick Bowe

seventy two. What U.S. Professional sports activities institution has been inside the equal town and with the identical call for the longest duration? Philadelphia Phillies

seventy three. In what sport would in all likelihood you word a bicycle kick? Soccer

seventy four. Having finished their home games in the Cycledrome, what end up the closing group to win an NFL Championship that is not inside the league? Providence Steam Roller

seventy five. To the nearest foot, how tall is the Green Monster at Fenway Park? 37'2"

76. Against what hitter did Nolan Ryan record his four,000th profession strikeout? Danny Heep

77. What U.S. City grow to be first of all provided the 1976 Winter Olympics? Denver, Colorado

seventy eight. How frequently did Babe Ruth win the World Series as a member of the New York Yankees? Four

seventy nine. What NFL group to begin with signed Kurt Warner as an undrafted loose agent? Green Bay Packers

80. What British Open winner's autobiography is titled, All My Exes Wear Rolexes? John Daly

eighty one. What changed into the primary European united states of america of america of the us to win the FIFA World Cup? Italy (1934)

eighty . What NBA participant earned the nickname, "Vinsanity"? Vince Carter

eighty three. Who have become the first MLB participant to earn $10 million dollars in a unmarried season? Albert Belle

84. Who managed Leicester City to the 2016-17 English Premier League perceive? Claudio Ranieri

80 five. Who turn out to be the number one primary pick in the 1975 NBA Draft, and the first defend to achieve 70 elements in an NBA enterprise? David Thompson

86. Who have end up the primary train to win hockey's "Triple Gold Club" (Stanley Cup, Olympic Games gold medal and World Championship gold medal)? Mike Babcock

87. Who headbutted Marco Materazzi at the 2006 FIFA World Cup? Zinedine Zidane

88. In what recreation should you find out umbrella and mushroom positional set ups? Water polo

Chapter 9: Partner Games

The following video games can be finished as character video games or may be executed collectively. I even have used numerous of these video games in my non-public education periods with families.

Cartwheel races

Equipment Needed: None decide the begin/finish line.

The object of this race is to cartwheel faster than the opportunity game enthusiasts. Great exercising!

Forward Roll/Backward Roll Races

Equipment Needed: None, decide start/give up line

The object of this venture is to out roll your accomplice, or set the timer and see which participant wins.

Consecutive Jumps

Equipment Needed: None decide the begin/prevent line.

The item of this exercise is to finish the soar faster than the other participant. I love this game because of the truth it's miles a calorie burner and a leg strengthener.

Select the style of jump and permit the race start. Jumps: Split soar, popcorn bounce (contact the floor and soar without delay up in the air), long soar, 2 jogging steps and soar, Jumping jack leap, one leg hop, jump backwards, jump sideways.

Newspaper Race

Equipment Needed: Two newspapers in line with participant, determine begin/surrender line

The object of this exercise is to pass the give up line first the usage of only newspapers to step on. Great game for aerobic and middle strengthening.

Two newspaper sheets required consistent with player. Each participant have to race to the turning point and once more, stepping simplest on his newspapers. He steps on one, lays the possibility in front of him steps on it, actions the primary ahead, and steps on it and so forth.

Balloon Belly Relay Race

Equipment Needed: A balloon for each organisation.

The object of this recreation is to coordinate on foot sideways to the end line collectively along with your associate. I like this hobby because it calls for the companions to run sideways within the path of the quit line, tough their coordination skills and once more it's miles a extremely good for teamwork abilties constructing.

To play this sport, use inflated balloons. In businesses of , participants run to the stop line, protecting the balloon among their bellies, requiring them to run sideways in the direction of the give up line. They cannot contact the balloon with every other body element. If a player drops the balloon he can pick out it up along with his hands to reposition it. Once this group finishes they want to run yet again and tag the subsequent players, ensure handy them the balloon. The first crew to complete wins.

Balloon Soccer

Equipment Needed: One balloon consistent with partnership

The object of this enterprise is to exercising football abilities using a balloon. This interest is incredible in your more younger gamers and Grandma too. Practice passing, throwing and kicking the balloon into a cause.

Solar Balloon

Equipment Needed: Beach ball or balloon

The item of this recreation is to keep a seashore ball or balloon in the air for a selected wide form of hits without letting it hit the ground. I like this sport because of the reality you may be as creative as you want, try to maintain the balloon inside the air with only the use of your knee or your head or elbow.

Balloon Finger Balance

Equipment Needed: One balloon consistent with player

The item of this sport is to balance a balloon on the surrender of your finger for the longest time frame.

The balloon must now not be held, best balanced, and it need to no longer be tapped. The finger need to be in direct contact with the balloon continually. This sport is proper for consciousness, attention and physical movement.

Don't Pop Or Else...

Equipment Needed: Place one message inner severa inflated balloons

The item of this interest is to sit down down at the balloon without breaking the balloon. If the balloon bursts, the participant has to perform a little aspect is written on the piece of paper this is inside the balloon.

Messages can be: 10 push-ups, 25 jumping jacks, I even have used this sport to get my children to assist me with circle of relatives chores, I could write, easy your room, wash the auto, fold laundry... cause them to humorous messages!

Sign Language

Equipment Needed: Pieces of paper and a pencil.

The item of this sport is to exercising your legs. I like this workout because it strengthens the middle muscle groups and legs of each participant, make sure to preserve the phrases brief.

Each player writes several terms on a piece of paper, the paper is then positioned right into a bag and every player ought to take a phrase.

When a word is chosen the player want to spell out the word on the paper chosen the usage of only his legs. The unique game enthusiasts attempt to guess the phrase.

Football Throws

Equipment Needed: Target, tire or football internet or purpose and a soccer

The object of this interest is to exercise eye hand coordination. I like this sport because you may adjust the goal as a gamers capability stage improves.

Set up your purpose and rely the quantity of times you could throw the soccer inside the goal.

Football Runs

Equipment Needed: Masking or vinyl tape

The object of this game is to enhance the agility of a player, and its a laugh.

Find an open floor place or flat place outside. Cut or tear off five portions of tape in 5 foot lengths. Make five parallel lines about yards, 3 yards, four yards and 5 yards apart. Run from the primary line to the second line and run lower lower back to the start. Run from the number one line to the 0.33 line and run again. Run from the primary line to the fourth line and run decrease again. Finally, run from the primary line to the 5th line and run again. Each time you get to a line, you want to the touch it with every arms. To upload to the assignment, you may do one or extra push-united stateswhen you get to every line.

Penny Pitching

Equipment Needed: Pennies, beads, or small rocks, a wall and chalk

The object of this sport is to task hand-eye coordination abilities. This is a fantastic recreation for certainly every body, collectively with Grandpa.

Start this exercise by using drawing limitations subsequent to a wall and some distance from it, approximately four toes apart and six ft prolonged. Draw a few one of a kind line at the top of the gambling area. To begin the sport, the primary participant stands within the returned of the line and throws a penny towards the wall. The next participant throws his or her penny toward the wall and attempts to hit the alternative player's penny. If she or he succeeds, that player takes each pennies. The participant who has clearly gained pitches the following penny. The game is sustained till one participant in the end finally ends up with all the pennies.

Rebound Marbles

Equipment Needed: Marbles

The item of this game is to hit a goal, I like this recreation because it calls for finesse, sincerely the proper touch.

This activity requires or more gamers and a flat floor close to a smooth wall. Each participant throws one marble in the direction of the wall simply so it bounces off and lands on the flat ground. These marbles grow to be the motive marbles. Players take turns taking pictures marbles one after the other within the direction of the wall, trying to rebound each marble to hit a intention marble. A marble that doesn't hit a purpose marble turns into each different intention marble. The first participant to hit any cause marble wins all of the reason marbles.

Balloon Partnership

Equipment Needed: Beach ball or balloon

The item of this recreation is to maintain a beach ball or balloon in the air for a centered sort of hits with out letting it hit the ground.

Additionally, one person can not contact the object times in a row. Set a cause collectively together with your associate for the variety of hits that you could make as a partnership.

One Leg Challenge

Equipment Needed: None

The item of this recreation is balance. Ask all game enthusiasts to face on one leg. This is an terrific evaluation of ankle energy, balance, and intellectual centeredness.

Variations: Hop on one leg, stand on one leg coping with your partner and region fingers on every other's shoulders, gently try to push your opponent backward and forward to mission their stability.

Mirror Image

Equipment Needed: None

The item of this sport is to reflect the moves and moves of the other individual.

Body movement wearing sports may be maximum revealing, confronting and worthwhile. Make your actions exciting and slow sufficient for the opportunity person to mime as even though they had been a complete duration replicate.

Stomp the Snake

Equipment Needed: ropes

The item of this activity is to chase the snake and step on him.

Choose a player to be the snake. The snake receives a chunk of rope. When the leader says "move", the snake will run preserving the rope with 2 hands within the again of them as they run. The other player(s) try and stomp at the snake by way of using way of jumping at the piece of rope. If your accomplice stomps to your rope they may be the brand new snake and get to run.

Basketball Dribble

Equipment Needed: Make an obstacle route outside the use of out of doors furnishings, small rubbish cans, or floor mats and you may need a basketball.

The object of this challenge is to project your dribbling competencies.

Set up your path, ensure there's area amongst them. If indoors, easy a space throughout the furnishings or assemble your obstacle course within the storage. Dribble the basketball across the barriers, adjusting them to suit your functionality and health diploma. Try dribbling with the alternative hand or alternating hands. Also strive dribbling low and dribbling at waist pinnacle. You might also additionally even strive some of the frilly moves you have a look at the professional basketball gamers use on the equal time as dribbling.

Wall Ball

Equipment Needed: Tennis ball and wall

The object of this sport is to challenge your hand eye coordination.

The concept is to throw and lure the ball steady with a series of obligations. When you pass over, you lose your turn and should begin all around the subsequent time. Here are some responsibilities:

Meensies: Throw the ball and seize it before it bounces—10 instances

Onesies: Throw the ball and capture it after one leap—9 times

Clapsies: Throw the ball and clap arms earlier than you entice it—8 instances

Kneesies: Throw the ball and make contact with your knees earlier than catching it—7 instances

Twosies: Throw the ball and trap it after bounces—6 instances

Under the knee: Lift your leg and throw the ball underneath the knee in advance than catching it—5 times

Highsies: Throw the ball as immoderate as you can in advance than catching it—four times

Touch the floor: Throw the ball and capture it after touching the floor—2 times

Turn round: Throw the ball and seize it after turning sincerely around—1 time

Note: After going through the collection without lacking, undergo over again the usage of the proper hand only, then with the left hand handiest.

Beanbag Race

Equipment Needed: Tape or string, one beanbag steady with player

The item of this game is to balance and tempo.

Mark strains ten toes aside on the ground with tape or string. Give each participant a beanbag and have the game fanatics stand on the beginning line. Announce, "Ready, set, pass!" and function every participant race toward the quit line in one of the following strategies: Crawling, on the same time as balancing beanbags on their backs Running, whilst squeezing beanbags between their knees. A player is disqualified if his beanbag falls before he reaches the give up line.

Chapter 10: Family Golf

Equipment Needed: Shovel, empty, clean tin can, golf putter (or little one's golfing membership, slender length of timber, or heavy cardboard rolled and taped proper right into a "membership"), numerous golfing balls, paper and pencil.

This is so clean to make in your outdoor! Keep it there so kids can exercise on their personal.

Dig a hole within the floor and place the can in it so the top of the can is flush with the ground. Mark a niche about ten ft some distance from the can. This can be the point from which gamers will putt. Give the balls to the first participant and project him to putt them into the can. When he has putted all of the balls, retrieve them from the can and file his score. A ball that lands inside the can ratings factors, and the ball nearest the can rankings one element. (This way, each player constantly scores at the least one factor.)Each participant takes a flip. Set a time restrict. The participant with the most points on the same time as time runs out is the winner. A infant playing on my own can strive to build up a positive kind of factors internal a given term.

Balloon Ping Pong

Equipment Needed: One balloon in line with player, desk tennis paddles or tennis racquet

The item of this endeavor is to exercising your desk tennis or tennis skills the usage of a balloon.

Practice serving, forehand and backhand with a companion.

Bombers

Equipment Needed: Marbles

The object of this challenge is to collect as quite a few your combatants marbles as viable. This is a first rate undertaking to take with you everywhere you bypass.

The first participant shoots a marble any distance. This marble turns into the goal marble. The second player stands over the target marble. Holding a marble at eye level, he attempts to drop his marble onto the intention marble. If he hits it, he wins the target marble. If he misses it, the first participant wins the second one participant's marble. Have the gamers take turns taking pictures the purpose marble and looking to hit it.

Hopscotch

Equipment Needed: Chalk

The object of this sport is to complete the complete series of hops.

Use chalk to attract hopscotch court docket docket, there can be no proper or wrong manner to attract a hopscotch courtroom docket, be creative! The first player hops up the court docket docket and again all over again, hopping in each location every up and decrease lower back. On the number one journey, he hops on his proper foot. On the second revel in, he hops on his left foot. On his zero.33 enjoy, he hops on alternating toes.

On the fourth adventure, he hops along along with his ft together. The player hops on this collection till he makes a mistake like hopping on a line, placing every toes down whilst he's alleged to be hopping on one foot, or hopping on the wrong foot, at which point he fouls out. The special game enthusiasts take turns hopping in the equal manner. When it's the number one player's flip over again, he begins hopping from in which he fouled out on his final flip. The winner is the primary participant to complete the entire collection of hops.

Blanket Juggle

Equipment Needed: Ball, blanket or towels or every

The object of this sport is to try to juggle balls as a partnership, it is a notable higher body workout too.

Take one towel consistent with couple and vicinity one or more balls within the towel. Without letting bypass of the towel, toss the towel up making the balls pop off of the towel. Count the range of times you can hold the balls doping up without losing a ball.

Push Wars

Equipment Needed: None

The item of this sport is to attempt to tug your partner off his feet. This activity is also appropriate for muscle strengthening and to undertaking your stability.

Player tries to push every different player however can most effective make touch with palms. Two human beings stand an palms-period apart (can truely put fingers on one-of-a-kind people shoulders to ensure of the distance). Players ought to maintain their private feet collectively. Then gamers located their arms in the the front of them, arms dealing with out. The two game enthusiasts then attempt to make the alternative participant lose their balance (i.E. Move their feet). The excellent element game enthusiasts are allowed to touch but the specific player's palms. It is turns into a struggle among pushing and now not pushing on the right time that allows you to either push your partner over or, have your associate lunge forwards and fall forwards.

He Ain't Heavy He's my Brother

Equipment Needed: Blanket or bed sheet, prevent watch

The object of this exercise is to drag your "brother" in a blanket in the shortest quantity of time.

Each player has a blanket, his companion will sit down down down on the blanket. On the command of "pass" the participant pulling the blanket will run as rapid as he can pulling his partner inside the blanket. The individual with the shortest time wins.

Supporting Squats

Equipment Needed: None

The item of this recreation is to rise up once more to back along with your associate. This is a terrific manner to construct leg electricity too.

Each associate will stand returned to again, slowly decrease your self together collectively with your companion into a squat function and return to a standing function with out losing lower back to decrease returned contact. Record the amount of times you're capable of perform this motion efficaciously.

Push'em Back

Equipment Needed: None

The item of this game is to stand again to lower back along aspect your accomplice and push your accomplice off of his base. This is a outstanding manner to construct leg strength too.

Each partner will stand back to all over again, at the command of "move" each participant will use his leg strength to push his accomplice off of his base. The first player to efficaciously push his companion wins.

Toe Tapping

Equipment Needed: None

The item of this undertaking is to strive tough not to get your toes stomped.

A pair faces every extraordinary, preserving arms, now try and faucet every different's toes, on the same time as simultaneously seeking to avoid having their toes tapped. Assure that gamers are similarly armed, bare foot to bare foot, or shoe to shoe. Once a player has had their feet tapped three times, they alternate companions with the dropping participant of every other pair.

Small Groups: three-6 Players

Small businesses are taken into consideration three -6 gamers. These video games provide a fantastic workout and masses of guffaws. Every night time can be Family Game Night!

I Love You

Equipment Needed: Chairs

This is a incredible exercise for own family people! The item of the game is to get your accomplice to grin.

Players sit down in a circle in chairs. Choose one player to be "it." The determined on participant sits on the lap of some different player and attempts to get him to grin. If the chosen participant can get him to mention, "Honey I love you but I virtually can't smile," with out smiling, she actions immediately to a few different participant and attempts once more. However, if she succeeds in making that participant smile, that participant will become the subsequent one to be "it."

Chocolate Challenge

Equipment Needed: Small candy in wrappers, cube and mittens

The object of the sport is to look how quickly a player can open up a candy wrapper with gloves.

Place a wrapped small chocolate bar in the middle of a table with more than one mittens next to it. The game enthusiasts take turns rolling multiple cube. When a player rolls a six, he has 30 seconds to put on the mittens, unwrap the candy bar and devour it, if the player can't open the bar interior 30 seconds, he can't consume the chocolate. Game maintains.

Newspaper Race

Equipment Needed: Two newspapers in line with player, decide start/end line

The object of this recreation is to move the end line first the use of simplest newspapers to step on. Great activity for cardio and middle strengthening.

Two newspaper sheets required in step with player. Each player want to race to the turning element and again, stepping satisfactory on his newspapers. He steps on one, lays the alternative in the the the front of him steps on it, actions the number one ahead, and steps on it and so on.

Zip Bong

Equipment Needed: None

The object of this endeavor is to dispose of gamers concerning humorous noises and now not displaying one's tooth. Just considering this recreation makes me chortle, wait till your own family plays this exercise, its hilarious! Oh with the resource of the manner, laughter burns electricity!

Players stand or take a seat in a circle. One player begins offevolved the game by means of the usage of pronouncing "zip". The starting participant specifies the path she or he is sending the zip. The individual subsequent to the starting participant then says "zip", as does the following participant, and so on.

Chapter 11: Cartwheel Races

Equipment Needed: None, determine the begin/give up line.

The item of this race is to cartwheel quicker than the alternative game enthusiasts. Great exercising!

Spud

Equipment Needed: Ball

The item of this enterprise is to react rapid to trap the ball. I love this sport because it does require all gamers to pay interest and to react rapid.

Give every participant diverse. Throw the ball up inside the air and call quite a number of. If the player catches the ball he calls a few one of a kind quantity. If he does not seize the ball he calls "FREEZE" and takes 3 large steps and attempts to hit a person with the ball beneath the waist. If he hits the man or woman then they get a element in opposition to them and get to name a current day quantity. If the participant misses the man or woman he is making an attempt to hit then he receives a point in competition to him and has to throw the ball up within the air.

Snake inside the Grass

Equipment Needed: None

The object of this recreation is to tag as many gamers feasible, I like this activity due to the fact all gamers get an exquisite exercising, specifically the snake. This sport is awesome if mom or dad is the snake. A phrase of warning, please remind all runners to not step at the snake.

One man or woman is the snake, who lies on the floor on his stomach. Everyone else gathers fearlessly round to the touch him. When the referee shouts "snake-in-the-grass" virtually without a doubt everybody runs, staying inside the bounds of the snake location, while the snake, moving on his belly, attempts to tag as many as he can. Those touched come to be snakes. The remaining person stuck is the snake starter inside the subsequent recreation. Make the steady vicinity pretty small.

Three-Legged Race

Equipment Needed: Ribbon to tie a partners leg. Determine starting and finishing strains

The object of this exercise is to run faster than the other gamers using most effective three legs. I use this interest as an exercising in my training to educate coordination, teamwork and capability.

Allow all gamers to practice this capability earlier than using it in a race. Form companies of partners. Each player must tie truely one in all his or her legs to the other player's leg sincerely so when they pass, they bypass with three legs. The purpose is to race from the vicinity to start to the reason line and once more another time. The first set of partners to acquire the location to begin all another time is the winner. You can fall and get again up and hold the race, but you could now not win except you capture up to folks that are regardless of the fact that shifting at a regular pace.

Sack Races

Equipment Needed: Pillowcases or burlap sacks. Determine start/quit lines

The item of this activity is to hop faster than your fighters, I like this sport due to the fact it's far superb for cardiovascular workout and leg strengthening and children love this recreation.

Each teen climbs into the sack and, keeping the rims of the sack up spherical his or her hips or waist, have to hop from the area to start to the purpose line and back yet again. Many kids will fall over and you may snort masses inside the route of this loopy game. If adults take maintain of a sack and start hopping with them, the children will squeal with laughter—the adults will look even sillier than the kids!

Clothespin Tag

Equipment Needed: Clothespins

The item of this game is to avoid being pinned thru others. I like this due to the fact it's far a exceptional cardiovascular workout and it's simply simple amusing.

Give every player six clothespins. The item is for every player to pin a clothespin on six remarkable humans, whilst avoiding being pinned via unique players. The winner is the primary individual to do away with all his clothespins.

Spud on a Spoon

Equipment Needed: plastic spoons and a potato steady with player. (I actually have substantially utilized tennis balls) Determine the begin/quit line

The object of this interest is to balance a potato on a spoon and flow as speedy as you can. I like this sport because it teaches manipulate and coordination and pace.

Everyone balances a potato on his or her spoon and at the same time as the whistle blows, you need to make it to the reason line and again with out dropping your potato. If you are making it decrease lower lower back first, you are the winner.

Hands Up

Equipment Needed: One ball and determine the gambling place.

The object of this exercise is to get all game enthusiasts to run spherical with their hands up. I love this endeavor because it's a super cardio exercise and it's hard to preserve your fingers above your head as you run.

Select one individual to be "IT". "IT" has the ball and runs to hit the alternative game enthusiasts , if a player is hit and all over again it need to be beneath the waist or it does no longer bear in thoughts then that player stays in the sport but now has to run round along together along with his fingers stretched above his head. The exercise continues till all game enthusiasts have their hands up.

Sock Wars

Equipment Needed: Socks

The object of this game is to try to take off exceptional game enthusiasts' socks

Goal is for game enthusiasts to get the socks off extraordinary gamers at the same time as keeping their very very own socks on. Players need to stay interior a predefined area. If a player loses both his/her socks, that player is removed. Players might not touch their personal socks (might not placed their personal socks decrease lower again on or pull up a 1/2 taken off sock). Players might not deal with other game enthusiasts. The very last participant with a sock closing wins.(Tip: If the game is going too extended, get rid of all gamers who do no longer have each socks on.)

Biggest Winner

Equipment: Cones for boundaries (if gambling outside) and dots for a "spoil region."

The item of the game is to get gamers to run. Choose one character to be the tagger. I like this sport as it calls for gamers to work as a crew.

This individual will run and attempt to tag the opportunity gamers. If tagged, gamers might be part of palms with the tagger. Together they will attempt to tag others. Taggers must speak and cooperate with each unique to decide what course they'll pass in. If the taggers release their palms they may not tag certainly absolutely everyone until they be a part of fingers all all over again.

Hug-a-tree

Equipment Needed: Blindfold and boundaries

The object of this hobby is to take a look at guidance from your companion and efficaciously perform the assignment. I like this recreation as it teaches game enthusiasts to examine course and to recall of the mission they're performing.

This recreation calls for 2 gamers and one blindfold or bandana. The gamers need to be vintage enough to lead every other blindfolded, and you may want to start with some guiding pointers—to move slowly, preserve the blindfolded man or woman securely through the arm, and use phrases to guide them over roots and extraordinary boundaries.

One participant is blindfolded and gently spun round 3 times until they lose their bearings. The unique participant then leads the blindfolded participant thru the out of doors, or woods taking a wandering course so the blindfolded player is even greater stressed about wherein they may be. They need to live close by of the beginning area. The leader alternatives a tree and leads the blindfolded individual as an awful lot because it. They say, "This is your tree," and placed the blindfolded participant's palms on the trunk.

The blindfolded participant can take as long as they prefer mastering their tree, feeling its trunk, gaining knowledge of where the branches are, and sorting out the roots or other first rate skills. Once they're finished, the leader courses them returned to the beginning vicinity and gets rid of the blindfold. Then the participant can try and find out their tree!

Simon Says

Equipment Needed: None

The item of this hobby is to trick the possibility gamers.

Players form a line handling the leader, who plays any movement announcing Simon says do this. If He doesn't say "Simon says" earlier than an motion then everybody who imitates the movement is out of the game. Continue till one character is left.

Balloon Panic

Equipment Needed: Two to a few inflated balloons in step with individual and a stopwatch.

The item of this workout is to maintain the balloons from hitting the floor.

Each person has a balloon, with the rest in a nearby pile. Everyone starts offevolved bouncing their balloons within the air. Every 5 seconds, each other balloon is introduced. When a balloon hits the ground, time stops. The organization attempts to higher its report with a few unique strive.

Chapter 12: Andrew Catch The Balloon

Equipment Needed: Balloon

The item of this sport is to react brief at the identical time as your call is known as.

Stand in a circle. Toss a balloon within the air and speak to a person's name. That individual want to capture the balloon in advance than it touches the ground. If the man or woman succeeds he/she then tosses the balloon up and calls the following call.

Andrew Elbow

Equipment Needed: Balloon

An extension of Andrew Catch the Balloon. Now the balloon isn't stuck, but saved inside the air.

As well as calling out someone's call, moreover call out a body component which that man or woman has to apply to keep the balloon in the air until he/she calls some different person's name and body detail.

Feather Breath

Equipment Needed: Feather in line with man or woman

The object of this game is to keep a feather floating in the air using incredible your breath. I actually have considerably applied this recreation to exercising deep respiration.

Give each person a feather. On the command of "bypass" each person need to try to maintain their feather inside the air. Whoever can deep their feather within the air without allowing the feather to fall is the winner.

I Need Help

Equipment Needed: Many inflated balloons.

The object of this game is to hold the balloon inside the circle and to project coordination abilities.

Start off with anyone in a circle, going thru inwards, and hands at the back of again. The objective is for certainly everybody to be within the middle keeping all balloons afloat. Put among 0 and 3 balloons in people's hands in the lower back of their backs. Participants want to now not permit directly to others how many they have got. The chief starts offevolved through looking to maintain three balloons afloat in the center. When it turns into hard, the leader calls a person's call and says "Courtney, I need your help!" That character is available in with all their balloons and enables until it becomes tough and then they name "Kyle, I need your assist!" If a balloon falls on the floor, it ought to be picked up thru someone inside the middle and stored afloat. Everyone is a winner on this recreation.

Balloon Finger Balance

Equipment Needed: One balloon according to participant

The object of this undertaking is to balance a balloon on the stop of your finger for the longest time period.

The balloon need to no longer be held, simplest balanced, and it need to no longer be tapped. The finger ought to be in direct touch with the balloon always. This interest is proper for interest, awareness and physical motion.

www.ingramcontent.com/pod-product-compliance
Lightning Source LLC
Chambersburg PA
CBHW071441080526
44587CB00014B/1940